# Studies in Classification, Data Analysis, and Knowledge Organization

More information about this series at http://www.springer.com/series/1564

Francesco Mola · Claudio Conversano
Maurizio Vichi
Editors

# Classification, (Big) Data Analysis and Statistical Learning

Springer

*Editors*
Francesco Mola
Department of Business and Economics
University of Cagliari
Cagliari
Italy

Maurizio Vichi
Department of Statistical Sciences
Sapienza University of Rome
Rome
Italy

Claudio Conversano
Department of Business and Economics
University of Cagliari
Cagliari
Italy

ISSN 1431-8814 ISSN 2198-3321 (electronic)
Studies in Classification, Data Analysis, and Knowledge Organization
ISBN 978-3-319-55707-6 ISBN 978-3-319-55708-3 (eBook)
https://doi.org/10.1007/978-3-319-55708-3

Library of Congress Control Number: 2017962105



Printed on acid-free paper

This Springer imprint is published by the registered company Springer International Publishing AG part of Springer Nature
The registered company address is: Gewerbestrasse 11, 6330 Cham, Switzerland

# Preface

This volume comprises the revised versions of the selected papers presented during CLADAG 2015, the 10th Scientific Meeting of the Classification and Data Analysis Group of the Italian Statistical Society (SIS) that was held in Santa Margherita di Pula, Cagliari, Italy, during October 2015. The meeting took place under the auspices of the International Federation of Classification Societies (IFCS) and of the Italian Statistical Society (SIS). The local organizer was the Department of Business and Economics of the University of Cagliari, Italy.

During the event, advanced methodological researches in multivariate statistics with a special vocation in Data Analysis and Classification were promoted, and the interchange of ideas in these fields of research had strong support, as well as the dissemination of novel concepts, numerical methods, algorithms, computational, and applied results. The scientific program of the conference included three keynote lectures, an invited session, 10 specialized sessions, 15 solicited sessions, and 15 contributed sessions. All the specialized and solicited sessions were promoted by the members of the Scientific Program Committee.

The papers included in this book were submitted after the end of the conference and passed a double-blind revision process. The table of contents is organized basing on specific macro-topics nowadays characterizing different fields of the analysis of complex data structures and Big Data. Accordingly, the volume is organized into the following parts:

Part I—Big Data,
Part II—Social Networks,
Part III—Exploratory Data Analysis,
Part IV—Statistical Modeling,
Part V—Clustering and Classification,
Part VI—Time Series and Spatial Data, and
Part VII—Finance and Economics.

The editors would like to express their gratitude to all the young statisticians working for Local Organizing Committee (in particular Dr. Massimo Cannas, Dr. Giulia Contu, Dr. Luca Frigau, and Dr. Farideh Tavazoee) for their enthusiasm in supporting the organization of this event from the very beginning. Their hard work contributed relevantly to the success of the event.

Last but not least, we thank all authors and participants, without whom the conference would not have been possible.

Cagliari, Italy Francesco Mola
Cagliari, Italy Claudio Conversano
Rome, Italy Maurizio Vichi
November 2017

# Acknowledgements

The organizing committee of the 10th Scientific Meeting of the Classification and Data Analysis Group of the Italian Statistical Society (CLADAG) is grateful for the scientific support of members of the International Federation of Classification Societies (IFCS) and of the Italian Statistical Society (SIS). The successful organization of the event benefited from the support of local organizers from the Department of Business and Economics of the University of Cagliari (Italy) and from the financial support of Fondazione Banco di Sardegna. Authors of this book would like to express their very great appreciation to members of both organizations.

SIS -CLADAG
Classification and Data Analysis Group
of the Italian Statistical Society

ifcs

International Federation
of Classifican Societies

Società Italiana di Statistica

Università degli Studi di Cagliari

Fondazione Banco di Sardegna

# Contents

# Contributors

**Claudio Agostinelli** Dipartimento di Matematica, Università di Trento, Trento, Italy

**Casper J. Albers** Department of Psychometrics & Statistics, University of Groningen, Groningen, The Netherlands

**Sabina Augustyn** Department of Statistics, Cracow University of Economics, Cracow, Poland

**Mateusz Baryła** Department of Statistics, Cracow University of Economics, Cracow, Poland

**Sonia Bergamaschi** Department of Engineering "Enzo Ferrari", Università di Modena e Reggio Emilia, Modena, Italy

**Silvia Biffignandi** University of Bergamo, Bergamo, Italy

**Sergio Camiz** Dipartimento di Matematica, Sapienza Università di Roma, Rome, Italy

**Massimo Cannas** Dipartimento di Scienze Economiche ed Aziendali, Università di Cagliari, Cagliari, Italy

**Giovanni Casale** Department of Economics and Law, University of Cassino and Southern Lazio, Cassino, Italy

**Valeria Caviezel** Department of Management, Economics and Quantitative Methods, University of Bergamo, Bergamo, Italy

**Daniela Cocchi** Department of Statistical Sciences, Università Cattolica del Sacro Cuore, Milan, Italy

**Giulia Contu** Department of Economics and Business, University of Cagliari, Cagliari, Italy

**Marcella Corduas** Department of Political Sciences, University of Naples Federico II, Naples, Italy

**Antonella Costanzo** National Institute for the Evaluation of Education System – INVALSI, Roma, Italy

**Silvia Creta** Sapienza Università di Roma, Rome, Italy

**Chiara Dal Bianco** Department of Economics and Management, University of Padua, Padova, Italy

**Antonio D'Ambrosio** Department of Economics and Statistics, University of Naples Federico II, Napoli, Italy

**Maria E. De Giuli** Department of Economics and Management, University of Pavia, Pavia, Italy

**Renato De Leone** School of Science and Technology, Section of Mathematics, University of Camerino, Camerino, Italy

**Francesca Di Salvo** Dipartimento di Scienze Economiche, Aziendali e Statistiche, Università degli Studi di Palermo, Palermo, Italy

**Patrick Doreian** Faculty of Social Sciences, University of Ljubljana, Ljubljana, Slovenia; Department of Sociology, University of Pittsburgh, Pittsburgh, USA

**Carlo Drago** University of Rome 'Niccolo Cusano', Rome, Italy

**Fabrizio Durante** Dipartimento di Scienze dell'Economia, Università del Salento, Lecce, Italy

**Anna Maria Falzoni** Department of Management, Economics and Quantitative Methods, University of Bergamo, Bergamo, Italy

**Gianluca Frasso** Faculté des Sciences Sociales, University of Liége, Liége, Belgium

**Luca Frigau** Department of Economics and Business, University of Cagliari, Cagliari, Italy

**John C. Gower** Department of Mathematics & Statistics, The Open University, Milton Keynes, UK

**Simona Gozzo** Department of Political and Social Sciences, University of Catania, Catania, Italy

**Alessandro Greppi** Department of Economics and Management, University of Pavia, Pavia, Italy

**Leonardo Grilli** Department of Statistics, Computer Science, Applications 'G. Parenti' University of Florence, Florence, Italy

**Carmela Iorio** Department of Industrial Engineering, University of Naples Federico II, Napoli, Italy

**Henk A. L. Kiers** Department of Psychometrics & Statistics, University of Groningen, Groningen, The Netherlands

**Michele La Rocca** Department of Economics and Statistics, University of Salerno, Fisciano, Italy

**Caterina Liberati** DEMS, Universitá degli Studi di Milano-Bicocca, Milan, Italy

**Paolo Mariani** DEMS, Universitá degli Studi di Milano-Bicocca, Milan, Italy

**Valentina Minnetti** Department of Statistic Science, Faculty of Information Engineering, Informatics and Statistics, Sapienza University of Rome, Rome, Italy

**Francesco Mola** Dipartimento di Scienze Economiche ed Aziendali, Università di Cagliari, Cagliari, Italy

**Dennis M. Montagna** Department of Economics and Management, University of Pavia, Pavia, Italy

**Omar Paccagnella** Department of Statistical Sciences, University of Padua, Padova, Italy

**Lucia Paci** Department of Statistical Sciences, Università Cattolica del Sacro Cuore, Milan, Italy

**Giuseppe Pandolfo** Department of Industrial Engineering, University of Naples Federico II, Naples, Italy

**Agnese Panzera** Department of Statistics, Computer Science, Applications 'G. Parenti' University of Florence, Florence, Italy

**Roberta Pappadà** Department of Economics, Business, Mathematics and Statistics, University of Trieste, Trieste, Italy

**Barbara Pawełek** Department of Statistics, Cracow University of Economics, Cracow, Poland

**Alessandro Pirisinu** Department of Business and Economics, University of Cagliari, Cagliari, Italy

**Antonella Plaia** Dipartimento di Scienze Economiche, Aziendali e Statistiche, Università degli Studi di Palermo, Palermo, Italy

**Józef Pociecha** Department of Statistics, Cracow University of Economics, Cracow, Poland

**Giovanni C. Porzio** Department of Economics and Law, University of Cassino and Southern Lazio, Cassino, Italy

**Giancarlo Ragozini** Department of Political Sciences, University of Naples Federico II, Naples, Italy

**Carla Rampichini** Department of Statistics, Computer Science, Applications 'G. Parenti' University of Florence, Florence, Italy

**Renata Rotondi** CNR—Istituto di Matematica Applicata e Tecnologie Informatiche Enrico Magenes, Milano, Italy

**Mariantonietta Ruggieri** Dipartimento di Scienze Economiche, Aziendali e Statistiche, Università degli Studi di Palermo, Palermo, Italy

**Roberta Siciliano** Department of Industrial Engineering, University of Naples Federico II, Napoli, Italy

**Serena Signorelli** University of Bergamo, Bergamo, Italy

**Giovanni Simonini** Department of Engineering "Enzo Ferrari", Università di Modena e Reggio Emilia, Modena, Italy

**Emiliano Sironi** Dipartimento di Scienze Statistiche, Università Cattolica del Sacro Cuore, Milan, Italy

**Claudia Tarantola** Department of Economics and Management, University of Pavia, Pavia, Italy

**Venera Tomaselli** Department of Political and Social Sciences, University of Catania, Catania, Italy

**Nicola Torelli** Department of Economics, Business, Mathematics and Statistics, University of Trieste, Trieste, Italy

**Carlo Trivisano** Department of Statistical Sciences, Università Cattolica del Sacro Cuore, Milan, Italy

**Elisa Varini** CNR—Istituto di Matematica Applicata e Tecnologie Informatiche Enrico Magenes, Milano, Italy

**Beatrice Venturi** Department of Business and Economics, University of Cagliari, Cagliari, Italy

**Domenico Vistocco** Dipartimento di Economia e Giurisprudenza, Università degli Studi di Cassino e del Lazio Meridionale, Cassino, Frosinone, Italy

**Maria Prosperina Vitale** Department of Economics and Statistics, University of Salerno, Fisciano, Italy

**Sebastiano Vitali** Department of Probability and Mathematical Statistics, Faculty of Mathematics and Physics, Charles University, Prague, Czech Republic

**Adalbert F. X. Wilhelm** Jacobs University Bremen, Bremen, Germany

**Song Zhu** International ICT Doctorate School, Università di Modena e Reggio Emilia, Modena, Italy

# Part I
# Big Data

# From Big Data to Information: Statistical Issues Through a Case Study

**Serena Signorelli and Silvia Biffignandi**

**Abstract** The present paper gives a short overview of the use of Big Data for statistical purposes. The introduction of different classifications of Big Data highlights the problems that arise when trying to use them in a statistical way. After that, a small-scale case study is presented by critically highlighting problems and solutions arising out of the transition from Big Data to information; it combines Census data from the Italian NSI with a telecommunication provider dataset.

**Keywords** Big Data · Quality · Representativeness · Communication · Mobility

## 1 Introduction on Big Data

In the last few years, the term Big Data has been used in various fields, especially in statistics. Unfortunately, a precise definition of Big Data does not exist, as it is a general concept related to many disciplines and to a wide amount of different data. Big Data can be classified into [6] social media, personal data, sensor data, transactional data, and administrative data (there is a debate whether this category can be considered as Big Data). In some cases, survey data quickly collected using technical tools and contacting a large amount of units could be considered as Big Data. From a statistical point of view, a huge amount of data could be considered as a positive aspect for the information provided through the data collection. Big Data, as the term suggests, carry a great quantity of data but quality is a characteristic to look at before using them for statistical purposes.

S. Signorelli (✉) · S. Biffignandi
University of Bergamo, via dei Caniana, 2, 24127 Bergamo, Italy
e-mail: serena.signorelli@unibg.it

S. Biffignandi
e-mail: silvia.biffignandi@unibg.it



F. Mola et al. (eds.), *Classification, (Big) Data Analysis and Statistical Learning*,
Studies in Classification, Data Analysis, and Knowledge Organization,
https://doi.org/10.1007/978-3-319-55708-3_1

Big Data can be useful for two main different purposes:

1. operational: businesses use them to analyze their performance and to improve it. Databases for managerial purposes, no need to extend results to collectivity;
2. statistical: should provide statistical information, i.e., data which are representative of the whole target population and are of good quality.

We refer to this second purpose. The statistical context has some particular issues, as prior characteristics: quality and representativeness. Big Data, differently from traditional probability-based survey data, are not collected and designed to a specific statistical purpose, but are "harvested" as they are. Therefore, traditional statistical approaches (inference or modeling techniques) are not immediately applicable. Big Data could contain errors of different nature and they need appropriate error categorizations and statistical methods, still under study.
Another issue concerns their volatility and instability; data coming from social networks could become incomparable from one day to the next, due to the recurring changes that providers introduce. Moreover, transactional or administrative data could change their structure and the way they are collected for operational and efficiency reasons. Other problems concern big dimensionality. Big Data have a big dimensionality but they represent only a specific part of the general population, and big effort is necessary to make them representative to the whole population (representativeness: attempt to generalize the results to the target population). This task is not easy, as Big Data collected through a variety of formats could catch units or phenomena that differ from the units or phenomena that are not collected (i.e., considering mobile data, people who use smartphones could behave differently from people who do not use them). An attempt of generalization has been made by Elliott [4] who built pseudo-weights in order to combine probability and non-probability samples. Representativeness is only one aspect of the statistical issues relevant in order to obtain statistical socioeconomic indicators taking Big Data as a source.
The other main statistical issue is represented by quality. The risks of poor Big Data quality arise at three steps affecting:

i. initial data loading: in addition to the six classic data quality dimensions (completeness, conformity, consistency, accuracy, duplication, and integrity), relevancy of the specific Big Data as a data source has to be considered;
ii. application integration: various sources of data available and integration have to be done carefully. Rather critical point to be implemented, each source has its own quality characteristics and different sources have heterogeneous characteristics;
iii. data maintenance: agents like private businesses or public bodies provide Big Data; need to check the persistence of the characteristics and quality of data.

The data gathered using Big Data technology is much more vulnerable to statistical errors (non-sampling and sampling) than using traditional data sources. User entry errors, redundancy, corruption, noise accumulation, and uncorrelation of model covariates with the residual error are problems that affect the value of the data [5, 10]. Two solutions are recommended in literature in order to deal with quality issues: [6] suggests the introduction of a Total Error framework specific for Big Data, based

on the Total Survey Error framework that already exists [2]. Biemer [1] has created the Big Data process map that contains three phases: generation of data; extraction/transformation/loading in a homogeneous computing environment; and analysis, when data are converted to information. For each phase, he individuates which kind of errors arises. Dufty et al. [3] proposes a framework that aims at assessing the quality of the data at three stages: input, when the data are acquired; throughput, when the data get transformed, analyzed, or manipulated; and output, the phase of reporting. It is necessary to specify in detail the characteristics of the framework and to apply it to obtain socioeconomic statistical information. The framework focuses on specific quality requirements and challenges to use Big Data in Official Statistics. Big Data could represent an opportunity for Official Statistics [7]; in fact, they open to the possibility for *nowcasting* (the prediction of the present) and/or they represent a source of data to complement and extend microlevel and small area analysis, which potentially ensure comparability of phenomena across countries.
The paper focuses on a case study by pointing out an original overview and analysis of existing databases for the use of Big Data for statistical purposes. The aim is to verify whether this new data can bring information that are complementary and coherent with Official Statistics sources (origin/destination matrix in this case).

## 2 Case Study

The aim of our case study is trying to put in a unique interpretative framework one traditional statistical source and one typical kind of Big Data in order to evaluate some informative potentialities of this approach.
The case study is based on the use of two datasets:

- the 2011 ISTAT origin/destination matrix from the 15th Population and housing census that contains data on the number of persons that commute between municipalities—or inside the same municipality—classified by gender, mean of transportation, departure timeslot, and journey duration. The spatial aggregation is represented by Italian municipalities. The 15th Census was carried out on 9 October, 2011; the questions regarding commuting pattern referred to "last Wednesday" or a typical working/studying day.
- One of the 2014 1st Telecom Italia Big Data Challenge datasets. In particular, we use the one named "Telecommunications - MI to Provinces", which provides information regarding the level of interaction between the areas of the city of Milan and the Italian provinces. The level of interaction between an area of Milan and a province is given as a number; it represents the proportion of calls issued from the area of Milan to the provinces (and viceversa). For each area, the dataset contains two proportions: one representing the proportion of inflow telephone traffic and the another one representing the proportion of the outflow. The spatial aggregation

of the dataset are the Milano GRID squares[1] and the Italian provinces. The values are aggregated in timeslots of ten minutes.

Our analysis is limited to Lombardy region, divided into twelve administrative provinces. We build commuting patterns concerning the city of Milan and all provinces. According to the 2014 Report by the Italian Media Safeguards Authority (AGCOM), Telecom Italia is the market leader in mobile telecommunications with an amount of 33.2% in 2013 (the year our data refer to), of which 29.8% are residentials and 47.9% business. A good users' coverage is guaranteed. Obviously, the findings might hold for all residential users under the hypothesis that the consuming behavior does not differ with respect to the telephone provider.
At first, this work consists in the analysis of each dataset separately. We show the results regarding the outflow from the municipality of Milan to each of the Lombardy provinces, both of people (ISTAT dataset) and phone calls (Telecom Italia dataset). The results are presented as geographic maps created using CartoDB.[2] The twelve provinces are ranked considering the amount of the outflow and are then split into seven buckets and coloured from dark (first position) to pale green (last position).

## 2.1 *ISTAT Dataset*

Figure 1 shows the way in which the provinces appear after the ranking considering the general outflow of ISTAT dataset.[3]
We filter the results by commuting purpose: work or study; then we split the results into four departure timeslots, as asked into the Census question:

- timeslot 1: before 7.15,
- timeslot 2: from 7.15 to 8.14,
- timeslot 3: from 8.15 to 9.14, and
- timeslot 4: after 9.14.

The last filter that can be applied is the means of transportation used: car or other means of transport.

## 2.2 *Telecom Italia Dataset*

This dataset contains data of November and December 2013. We use only weekdays of the four complete weeks of November and compute the average over 20 days in

[1]Some of the datasets of the 1st Telecom Big Data Challenge referring to the Milano urban area are spatially aggregated using a grid composed by 10,000 cells over the municipality of Milan.
[2]https://cartodb.com/.
[3]The complete set of ranking maps can be found at the following link: http://www.data.unibg.it/dati/bacheca/90/79946.pdf.

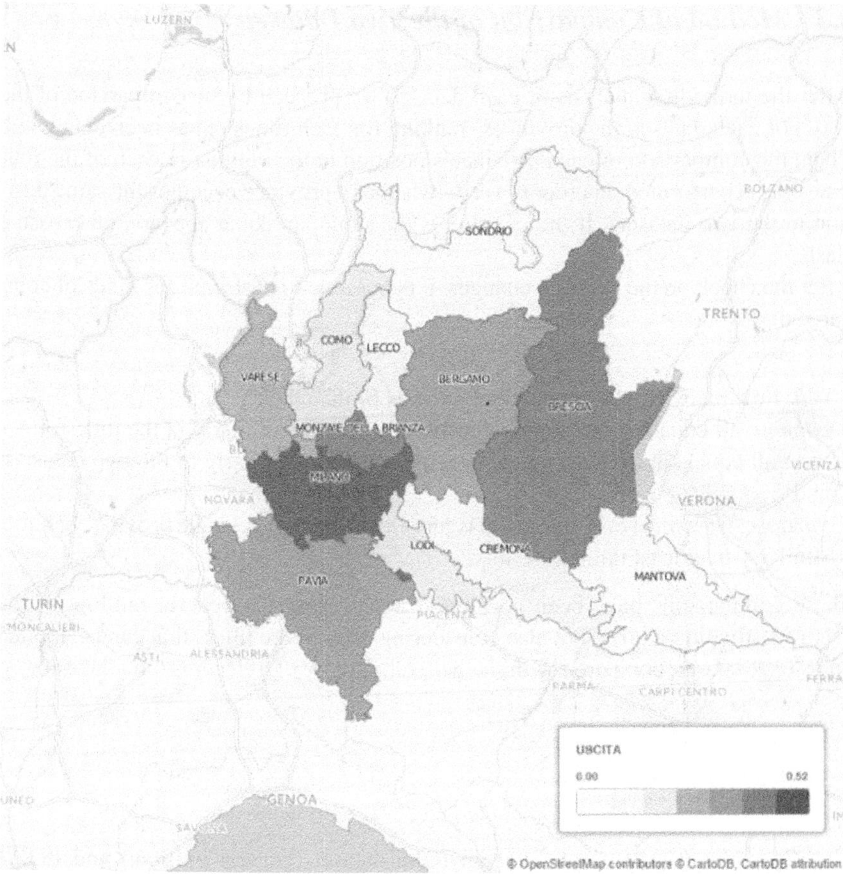


**Fig. 1** General outflow from Milan to provinces—ISTAT dataset

order to have a mean value comparable to the one from ISTAT.
It is not possible to split calls by commuting purpose. It is possible to split into timeslots as similar as possible to the ISTAT ones. Data are originally grouped into slots of ten minutes, so we build the following timeslots:

- timeslot 1: 6.20–7.10,
- timeslot 2: 7.20–8.10,
- timeslot 3: 8.20–9.10, and
- timeslot 4: 9.20–10.10.

We create the same kind of maps for the inflow from the Lombardy provinces to the city of Milan (Figures are provided into the link available at note 2).

## 2.3 Method of Comparison of the Two Datasets

After the individual analysis of each dataset, we proceed to the comparison of the two. For each dataset, the provinces' ranking for each timeslot has been computed. Then, the comparison of each province's position in the ranking in each of the four timeslots is performed, in order to verify whether a province occupies the same position in the two datasets. If the position is the same, an equal appears, otherwise a slash.
After the check on the position changes, it is possible to perform four kinds of comparison:

- *cell*: how many equals appear over 48 cells of the table;
- *column*: all equals in a column, all provinces behave similarly in the timeslot;
- *row*: all equals in a row, a single province behaves similarly in the two datasets; and
- *partial row*: at most only one equal is missing in one row, a single province behaves similarly in at least three timeslots.

These comparisons have been performed considering the general outflow of the origin/destination matrix, but also considering each of the filters that can be applied on ISTAT dataset, compared with the general outflow of Telecom Italia dataset.

## 2.4 Results

The comparison is between Telecom Italia dataset (general outflow) and ISTAT dataset (considering general outflow and filters shown in Sect. 2.1) broken down by timeslots. The four types of comparison (cell, column, row, and partial row) are performed.
Outflow results (in terms of number of matches) are shown in Table 1. A perfect match is found only in the general outflow. Rather many matches exist in the work outflow and in the outflow by car. A similar analysis with the same filters is performed for inflow. The results are presented in Table 2. The situation is very different, very few matches appear. Anyway, some matches arise in cell comparison, especially considering inflow by car, work inflow by car, and work inflow. The lack of matches could be influenced by the heavy traffic flow of people daily commuting to Milan.

The findings of this case study show some opportunities of the use of mobile Big Data, like providing increased information on the social exchanges in a physical and communication perspective between provinces. If the matching is satisfactory, Big Data could represent a source for Official Statistics, cheap and up-to-date. Furthermore, the number of matches that we adopted in our analysis is an interesting measure of the size of the flows and of the communications. Standardized values, i.e., with respect to the whole population (in the case of work commuting, with respect

**Table 1** Outflow results (number of matches)

| Telecom Italia versus ISTAT | Cell (48)[a] | Column (48)[a] | Row (12)[a] | Partial row (12)[a] |
|---|---|---|---|---|
| Outflow | 48 | 4 | 12 | 12 |
| Work outflow | 44 | 2 | 8 | 12 |
| Study outflow | 16 | 0 | 1 | 2 |
| Outflow by car | 40 | 0 | 7 | 6 |
| Outflow by other transport | 28 | 0 | 2 | 3 |
| Work outflow by car | 31 | 0 | 4 | 10 |
| Work outflow by other transport | 22 | 0 | 2 | 4 |

[a]maximum number of matches

**Table 2** Inflows results (number of matches)

| Telecom Italia versus ISTAT | Cell (48)[a] | Column (4)[a] | Row (12)[a] | Partial row (12)[a] |
|---|---|---|---|---|
| Inflow | 18 | 0 | 3 | 5 |
| Work inflow | 22 | 0 | 1 | 6 |
| Study inflow | 13 | 0 | 1 | 2 |
| Inflow by car | 23 | 0 | 1 | 4 |
| Inflow by other transport | 21 | 0 | 3 | 4 |
| Work inflow by car | 23 | 1[b] | 1 | 4 |
| Work inflow by other transport | 21 | 0 | 1 | 5 |

[a]maximum number of matches
[b]two missing values

to the work age range), could highlight other interesting aspects regarding communication and mobility behaviors. Moreover, the possibility to carry out an analysis of the phone calls that take into account also the user profile could give a remarkable knowledge contribution even though in full compliance of privacy issues. In the future, it could be evaluated integrating these data in Official Statistics. For instance, these data could be useful to create indicators of social exchange and interaction that could be used as a base for a proxy of the demographic mobility in the time interval of the Census data collection. Our study is a preliminary feasibility analysis; further analyses will be planned. It presents some limits:

- the two flows represent different purposes: Telecom Italia dataset does not specify a specific purpose, while ISTAT dataset only contains work and study flows;

- for a more extended users coverage, other providers could be considered. This could help in evaluating the stability of the behavior across users of different providers, i.e., in understanding if our findings could apply to target population;
- Telecom Italia dataset only contains traditional phone calls, other types of calls (i.e., Skype and Whatsapp) which are now very spread are not considered;
- Telecom dataset contains province reference, it would be useful to have municipalities to do a better match with ISTAT and to map commuting patterns.

## 3 Conclusions

Big Data potentialities for Official Statistics need a huge amount of experimentation and of economic statistics studies to set up a suitable metadata framework and to evaluate Veracity, Validity, and Value of the considered Big Data. Surely, deeper insight in quality of Big Data and in the variety of aspects and sources which could be integrated to set up their potential use as a statistical information is needed.
The overview of potentialities and problems presented in our paper highlights most critical research points and the present case study shows some innovative ideas on how to go through the tentative use of Big Data in Official Statistics. Some potentialities seem to be expected. In particular, our case study shows how mobile phone calls could be investigated with respect to mobility. Official Statistics data highlights how these data could catch jointly social communication and physical mobility aspects. Obviously, further research is needed; for instance, [a.] more detailed analyses on similarities and differences between the two datasets; [b.] the search for more possible data to be considered in the comparison and the identification of different case studies on Big Data analysis.

**Acknowledgements** The authors acknowledge financial support by the ex 60% University of Bergamo, Biffignandi grant.

## References

1. Biemer, P.P.: Dropping the 'S' from TSE: Applying the Paradigm to Big Data (2014). https://www.niss.org/sites/default/files/biemer_ITSEW2014_Presentation.pdf
2. Biemer, P.P.: Total survey error: design, implementation, and evaluation. Public Opin. Q. **74**(5), 817–848 (2010)
3. Dufty, D., Brard, H., Reedman, L., Lefranc, S., Signore, M., Munoz, J., Ordaz, E., Struijs, P., Maslankowski, J., MaRozkrut, D., Nikic, B., Jansen, R., Kovacs, K., Jug, M.: A Suggested Framework for National Statistical Offices for assessing the Quality of Big Data, Deliverables of the UNECE Big Data Quality Task Team (2014). https://ec.europa.eu/eurostat/
4. Elliott, M.R.: Combining data from probability and non-probability samples using pseudo-weights. Surv. Pract. **2**(6), 1–7 (2009)
5. Fan, J., Han, F., Liu, H.: Challenges of Big Data analysis. Natl. Sci. Rev. **1**(2), 293–314 (2014)

6. Japec, L., Kreuter, F., Berg, M., Biemer, P., Decker, P., Lampe, C., Lane, J., O'Neil, C., Usher, A.: Big Data in survey research: AAPOR task force report. Public Opin. Q. **79**(4), 839–880 (2015)
7. Kitchin, R.: The opportunities, challenges and risks of Big Data for official statistics. J. Int. Assoc. Official Stat. **31**, 471–481 (2015)
8. Laney, D.: 3D data management: controlling data volume, velocity, and variety. META Group Research Note, 6 (2001). http://blogs.gartner.com/doug-laney/files/
9. OECD: Quality Framework and Guidelines for OECD Statistical Activities, OECD (2012). http://www.oecd.org/std/qualityframeworkforoecdstatisticalactivities.htm
10. Saha, B., Srivastava, D.: Data quality: the other face of Big Data. In: International Conference on Data Engineering (ICDE), IEEE, pp. 1294–1297 (2014). http://ieeexplore.ieee.org/stamp/stamp.jsp?arnumber=6816764

# Enhancing Big Data Exploration with Faceted Browsing

**Sonia Bergamaschi, Giovanni Simonini and Song Zhu**

**Abstract** Big data analysis now drives nearly every aspect of modern society, from manufacturing and retail, through mobile and financial services, through the life sciences and physical sciences. The ability to continue to use big data to make new connections and discoveries will help to drive the breakthroughs of tomorrow. One of the most valuable means through which to make sense of big data, and thus make it more approachable to most people, is data visualization. Data visualization can guide decision-making and become a tool to convey information critical in all data analysis. However, to be actually actionable, data visualizations should contain the right amount of interactivity. They have to be well designed, easy to use, understandable, meaningful, and approachable. In this article, we present a new approach to visualize huge amount of data, based on a Bayesian suggestion algorithm and the widely used enterprise search platform Solr. We demonstrate how the proposed Bayesian suggestion algorithm became a key ingredient in a big data scenario, where generally a query can generate so many results that the user can be confused. Thus, the selection of the best results, together with the result path chosen by the user by means of multifaceted querying and faceted navigation, can be very useful.

**Keywords** Bayesian network · Faceted browsing · Big Data

---

S. Bergamaschi (✉) · G. Simonini
Department of Engineering "Enzo Ferrari",
Università di Modena e Reggio Emilia, Modena, Italy
e-mail: sonia.bergamaschi@unimore.it

G. Simonini
e-mail: giovanni.simonini@unimore.it

S. Zhu
International ICT Doctorate School,
Università di Modena e Reggio Emilia, Modena, Italy
e-mail: song.zhu@unimore.it



F. Mola et al. (eds.), *Classification, (Big) Data Analysis and Statistical Learning*,
Studies in Classification, Data Analysis, and Knowledge Organization,
https://doi.org/10.1007/978-3-319-55708-3_2

# 1 Introduction

With the modern information technologies, data availability is increasing at formidable speed giving rise to the Big Data challenge [1, 2]. As a matter of fact, Big Data analysis now drives every aspect of modern society, such as manufacturing, retail, financial services, etc. [3–5]. In this scenario, we need to rethink advanced and efficient *human–computer interaction* to be able to handling huge amount of data. In fact, one of the most valuable means to make sense of Big Data, to most people, is data visualization. As a matter of fact, data visualization may guide decision-making and become a powerful tool to convey information in all data analysis tasks. However, to be actually actionable, data visualization tools should allow the right amount of interactivity and to be easy to use, understandable, meaningful, and approachable.

In this article, we present a new approach to visualize and explore a huge amount of data. In particular, the novelty of our approach is to enhance the faceted browsing search in `Apache Solr`[1] (a widely used enterprise search platform) by exploiting Bayesian networks, supporting the user in the exploration of the data. We show how the proposed Bayesian suggestion algorithm [6, 7] be a key ingredient in a Big Data scenario, where a query can generate too many results that the user cannot handle. Our proposed solution aims to select best results, which together with the result path, chosen by the user by means of multifaceted querying and faceted navigation, can be a valuable support for both Big Data exploration and visualization.

A study about recommendation system for visualization is explored by Heckerman et al. [8], where they presented a study about a dependency network and demonstrated how it is useful for the visualization of acausal predictive relationships. Our approach employs Bayesian Network (similarly to dependency network), and investigate the exploitation of user feedback (i.e., navigation path) for faceted browsing.

The rest of the paper is organized as follows: Sect. 2 introduces the *faceted browsing* technique; Sect. 3 describes how it is enhanced exploiting Bayesian networks; Sect. 4 illustrates the technological architecture of our tool; Sect. 5 shows some preliminary tests; finally, conclusion and future work are presented in Sect. 6.

# 2 Faceted Browsing and Big Data

The faceted browsing [9] (or faceted navigation) is a technique offered by many search engines for accessing information. It allows to explore data applying dynamic filters in multiple steps: each time a filter is applied, the results are shown to the user, which can apply additional filters or modify existing ones.

An example of a faceted panel of an online store is shown Fig. 1. For each facet, there are one or more values, called the facet values, used as filter for refining search query, interactively. Moreover, a facet counter may be associated with each value representing the number of records matching with this value. For instance, consider

[1] http://lucene.apache.org/solr/resources.html.

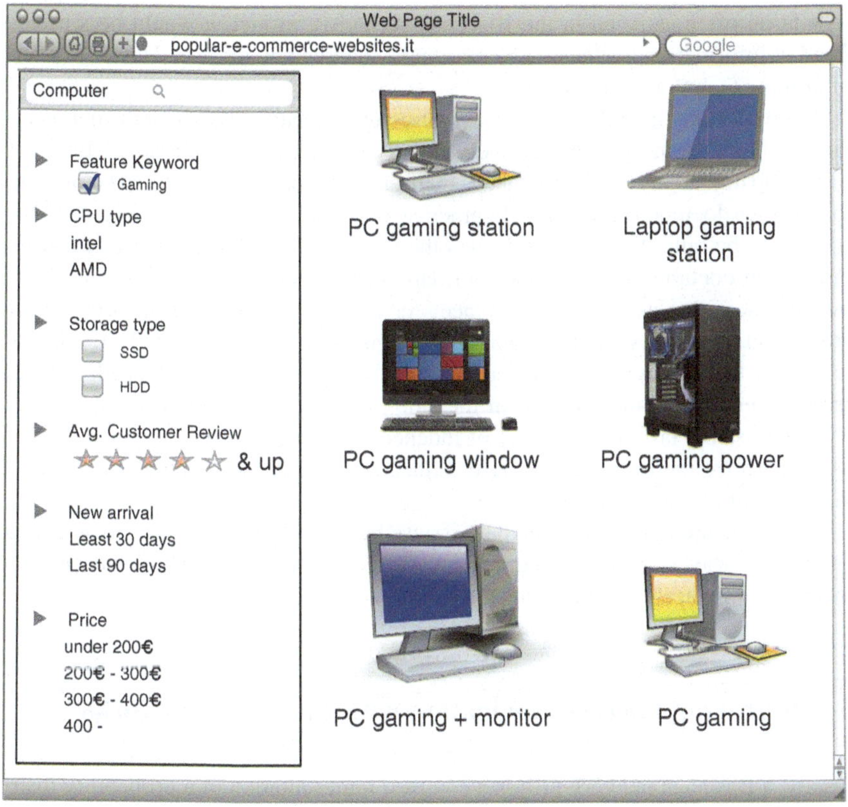


**Fig. 1** Faceted panel for product searching in an online store

Fig. 1, where facets are present in different forms. In this example, there are multiple "OR" choice (as Feature Keyword, Storage Type), single value choice (CPU Type), ranged value (Avg Customer Review, Price), and custom range input.

Typically, a parametric query for a faceted content collection is Boolean query that chains with the AND operator a series of OR constraints: values selected within a single facet are combined using a logical OR, whereas constraints associated with different facets are combined using a logical AND. The result is set of objects in the collection that satisfies it.

In contrast, the faceted navigation allows the user to elaborate a query progressively, seeing the effect of each choice inside one facet on the available choices in other facets. From the user's perspective, faceted navigation eliminates the "dead ends" that may result from selecting unsatisfiable combinations of constraints among the facets. In fact, most combinations of facet values are unsatisfiable, because the set of satisfiable combinations is typically a sparse subset of the set of all possible combinations. Furthermore, a search engine should not present to the user all facets and all facet values if the cardinality of the facet categories and their values are huge (condition often satisfied in the Big Data scenario); in fact, it would be a useless flood of information that cannot be reasonably handled by the user. Hence, the need of pruning techniques arises.

Our case study includes textual/semi-structured documents, where the number of facets reflects the number of ways a document can potentially be classified. In theory, there is no limit to the number of facets: there are infinitely many potential taxonomies to classify a document collection. In practice, of course, the number of facets is finite, but it may be quite large. There is also the issue of dependence among facets. For instance, if documents containing information about cities, states, and countries, we may devise either as three distinct facets or as a single hierarchical facet. On the other hand, languages and nationalities are highly correlated and yet clearly distinct facets. At best, designing a faceted classification scheme with independent facets requires an extraordinary effort on the side of information architects; at worst, it is an impossible task as such a set of independent facets would not match the way users conceive the information space. Either way, we cannot require that facets be independent of one another.

Our work attempts to address these issues, by exploiting a probabilistic graphical model (i.e., Bayesian network) to capture facets dependencies and to determine the most valuable facets to be presented to users.

## 3 Improving Faceted Navigation with Bayesian Networks

Briefly, a Bayesian network [10] is a compact representation of a probability distribution associated with a set of related variables. The network has two components: a probabilistic *Directed Acyclic Graph* (DAG called a structure) and a set of *Conditional Probability Tables* (CPTs). The DAG represents graphically the conditional dependencies between variables. The nodes of a structure correspond to the variables of interest and its edges have a formal interpretation in terms of probabilistic independence. A CPT is a table associated with a node (each node has its own CPT), where it is defined by the conditional probability of the single variable (represented by the node) with respect to the others (represented by other nodes connected to that node).

In a Bayesian network, a *finding* is an assignment of a value to a variable as a consequence of an observation. A set of finding is called an *evidence* case. Whenever an *evidence* related to a set of nodes occurs, the conditional probabilities of the neighbor nodes change and propagate changes through all the network.

An example is shown in Fig. 2, where we consider the relation between a disease and a test to diagnosis it. On the left side of the figure, the *Test* node is almost negative; then, after a finding is set (i.e., the *Test* is observed positive), the probabilities change as shown in the right side of the figure. The node *Test* is colored in gray to denote the existence of a finding, while the probabilities $P$ of the node *Disease* have changed.

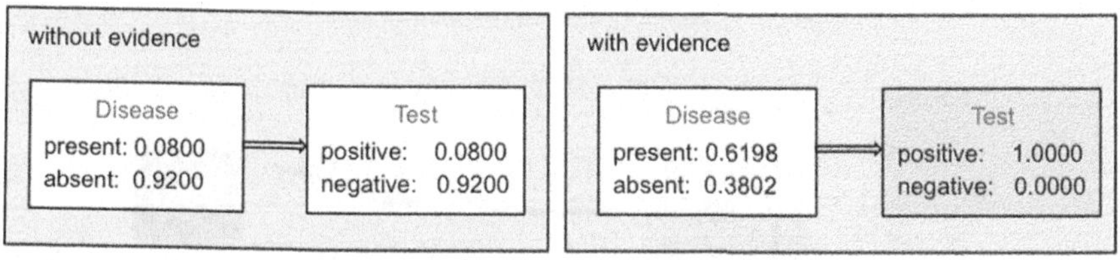


**Fig. 2** Bayesian network of disease and test

In our model, the variables modelled inside a Bayesian network are the attributes of a dataset, i.e., the facets. Thus, Bayesian networks can be exploited to infer the relationship among these facets. We develop our tool on top of `Apache Solr`, enhancing its faceted browsing interface, integrating facilities offered by `OpenMarkov`[2] to automatically learn a Bayesian network starting from data (e.g., from a `cvs` file). In our tool, user interactions with faceted navigation update evidences of the Bayesian Network's variable, and consequently this change influences the probability of the network. Thus, the tool shows how the search fields are dependent on each other in the form of graphs. The tool interface allows to give suggestion on the facets that she/he consider relevant to the graph, trying to find out other relevant features by using relationships among the attributes on the Bayesian networks. In order to limit the number of items in each facet, the system calculates two groups of similar and dissimilar items, ranks them, and returns a selection of the top $n$ items to the user: The similar items help the user to define precise search, while the dissimilar ones stretch the range of search. Another key feature of our tool is the query recommendation, which can guide the user in formulating queries. In fact, when the user hovers over a facet in the `Solr` selection panel, the interface communicates to the tool the current query and the facet that the user intends to select. Then, it returns suggestions on the basis of how that choice would change probabilities of other facets, accordingly to the Bayesian network. So, the user is facilitated in his request as she/he will have a real-time feedback on the effects of his/her search task.

## 4 Big Data and Solr

In a Big Data exploration context, a search engine with Big Data support is required, and `Apache Solr` is perfect as a basic platform in our tool. To enable a Big Data solution for `Solr`, we integrated it with other technologies: `Tomcat`, `Apache Zookeeper`, and `Hadoop Distributed File System(HDFS)`. `Solr` runs by default on servlet `Jetty`. However, for a Big Data tool a more powerful servlet is required. Thus, we chosen `Tomcat` because it is open and it performs better than `Jetty`.

[2]http://www.openmarkov.org/users.html.

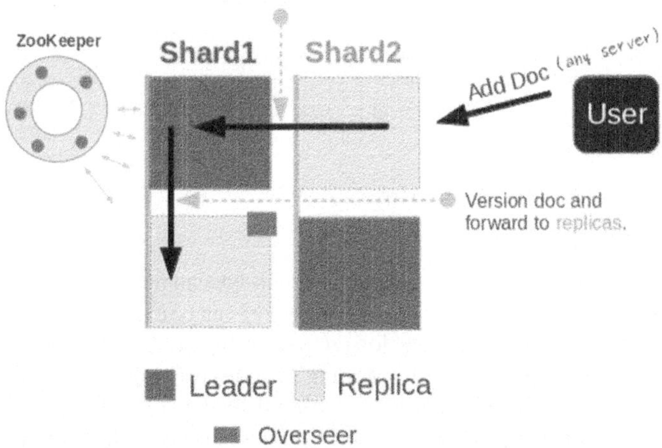


**Fig. 3** SolrCloud with Apache Zookeeper

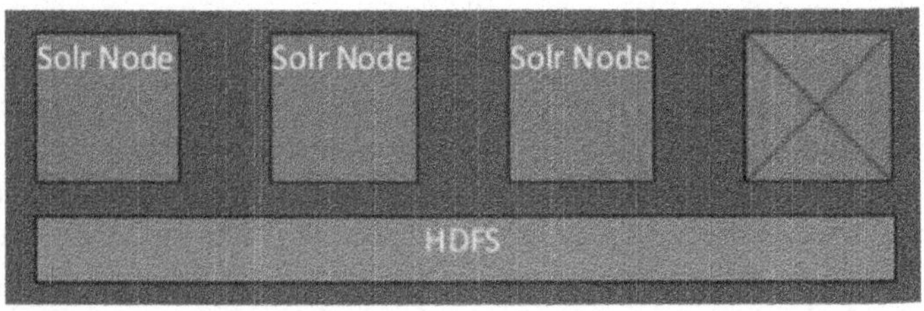


**Fig. 4** Solr and HDFS schema

To process Big Data, `Solr` must run in a multinode environment, which can ensure scalability and fault tolerance property. Consequently, the solution we devised is to exploit the distributed capabilities of `Solr` called *SolrCloud*. With this feature, a single search index can span on several nodes with `Solr` installed. Thus, the best way to manage a multinode `Solr` application is thought *sharding* and *replication* [11]. In detail, *sharding* is a horizontal partitioning that separates data into smaller partition called *shard*. In addition, *replication* is the process of creation and synchronization of replicas. In this architecture, the data is partitioned into shards and replicated. Moreover, the original shards are called leaders, and their replicas contain the same data, while the administrative tasks are managed by leaders.

In this context, `Apache Zookeeper` is used as repository for cluster configuration and coordination, and contains information about all `Solr` nodes (Fig. 3).

Finally, we configured `Solr` to use `HDFS` as repository for index and transaction log files (Fig. 4).

# 5 Test

We modified the `Solr` front-end equipping it with a new faceted search interface and integrating it with a customized `OpenMarkov` API [12, 13].

We tested our tool with a mushrooms dataset,[3] consisting of thousands of instances, each having 22 categorical attributes (e.g., cap shape, cap surface, cap color, etc.). The choice of the dataset is due mainly to its complexity (the deriving network of attributes is shown in Fig. 5). In fact, the goal is to define an approach capable to handle such level of complexity of schema, while the scalability is guaranteed the underlying HDFS and Lucene. Moreover, the mushrooms dataset is a well-known dataset where that yields results easy to interpret.

We employed the *Hill Climbing* algorithm [14] to automatically learn the Bayesian network. This algorithm performs a search by departing from a network without links and adding at each step the link leading to the highest score, provided that the score was positive. The advantage of this algorithm is that it performs a heuristic to automatically search through the space of possible structures, using a metric that measures how well each structure can represent the probability distribution of the variables of the dataset [6]. The result of the learning is indicated in Fig. 5 (Mushroom network).

To conclude, we present an example of the faceted browsing in our tool in Fig. 6. In the figure, we show how the interface appears to a user hovering the mouse over *brown* of the *CapColor* facet. The pop-up shows which and how facets are affected, and what facets may be proposed to the user on the basis of the learned Bayesian network. In detail, the *unaltered* column contains the facet values present in the current selection and that would remain unaffected (i.e., they remain in the facet tab) after the application of the *brown* filter; the *added* column contains the facet values

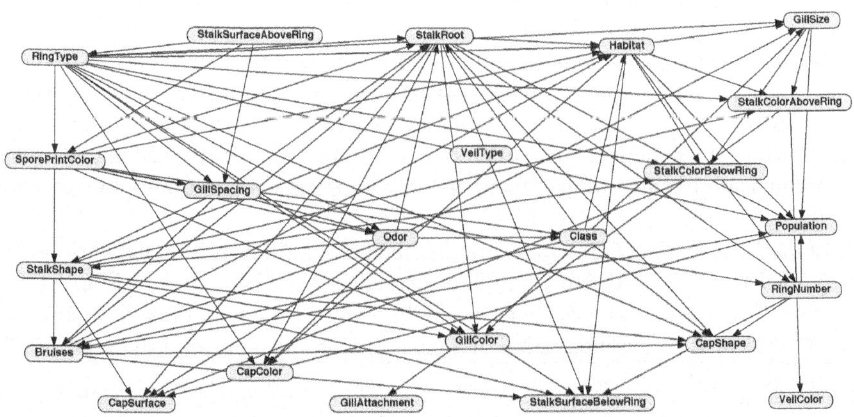


**Fig. 5** Mushroom network obtained from OpenMarkov

[3]http://archive.ics.uci.edu/ml/datasets/Mushroom.

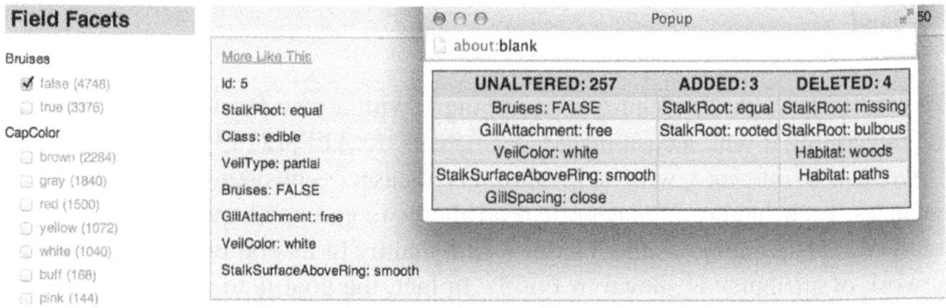


**Fig. 6** Faceted browsing and query advisor on Solr

not present in the current selection and that would be added in the facet tab after the application of the *brown* filter; the *deleted* column contains the facet values present in the current selection that would be deleted to the facet tab after the application of the *brown* filter.

## 6 Conclusion and Future Work

In the era of Big Data, the risk for the users of search engines is to be overwhelmed from data. Thus, a visualization tool enabling an intelligent exploration of the data is of paramount importance for the usability of such systems. In this paper, we present the preliminary results of our approach to the Big Data visualization challenge. Our proposal is a dynamic and visual tool for big data based on an extension of the popular open-source enterprise search engine `Apache Solr`. In particular, we propose the novel idea of enhancing the faced navigation feature by exploiting a probabilistic graphical *Bayesian Network* model.

Bayesian network learning algorithms have a high computation cost, especially when the data set is big. Thus, a scalability problem for a Big Data application, where the volume of data is huge, arises. In literature, when there is a scalability problem, one of the best solutions is parallel processing. In fact, numerous machine learning algorithms are rewritten for the `Apache Spark` framework in library like `MLlib` [15].

In the Bayesian network context, a distributed algorithm for the network learning called Consensus Monte Carlo is discussed in [16]. Moreover, a MapReduce approach for learning process is proposed by Fang et al. [17]. Thus, the next step of our work will be the integration of a parallel learning algorithm into our tool to assure the scalability issue.

**Acknowledgements** We would like to thank Paolo Malavolta and Emanuele Charalambis for working on this project for their master thesis as students of the DBGroup (www.dbgroup.unimo.it) of the University of Modena e Reggio Emilia, during their period abroad, hosted by the University of Michigan under the supervision of professor H. V. Jagadish.

## References


1. Bergamaschi, S.: Big Data analysis: Trends & challenges. In: Proceedings of 2014 International Conference on High Performance Computing & Simulation (HPCS), IEEE, pp. 303–304 (2014)
2. Bergamaschi, S., Ferrari, D., Guerra, F., Simonini, G., Velegrakis, Y.: Providing insight into data source topics. J. Data Seman. **5**(4), 211–228 (2016)
3. Labrinidis, A., Jagadish, H.V.: Challenges and opportunities with Big Data. Proc. VLDB Endowment **5**(12), 2032–2033 (2012)
4. Simonini, G., Bergamaschi, S., Jagadish, H.V.: BLAST: a loosely schema-aware meta-blocking approach for entity resolution. PVLDB **9**(12), 1173–1184 (2016)
5. Guerra, F., Simonini, G., Vincini, M.: Supporting image search with tag clouds: a preliminary approach. Adv. Multimedia **2015**, 1–10 (2015). https://doi.org/10.1155/2015/439020
6. Cooper, G.F., Herskovits, E.: A Bayesian method for constructing Bayesian belief networks from databases (2013). arXiv:1303.5714
7. Simonini, G., Song, Z.: Big Data exploration with faceted browsing, International Conference on High Performance Computing & Simulation, HPCS 2015, IEEE, pp. 541–544 (2015)
8. Heckerman, D., Chickering, D.M., Meek, C., Rounthwaite, R., Kadie, C.: Dependency networks for inference, collaborative filtering, and data visualization. J. Mach. Learn. Res. **1**, 49–75 (2000)
9. Yee, K.P., Swearingen, K.L., Li, Hearst, M.: Faceted metadata for image search and browsing. In: Proceedings of the SIGCHI Conference on Human Factors in Computing Systems (CHI '03). ACM, New York, NY, USA, pp. 401–408 (2003)
10. Nielsen, T.D., Jensen, F.V.: Bayesian Networks and Decision Graphs. Springer Science & Business Media, Berlin (2007)
11. Grainger, T., Potter, T., Seeley, Y.: Solr in action, Manning (2014). www.manning.com
12. Malavolta, P.: Faceted browsing: analysis and implementation of a Big Data solution using Apache Solr (2014). https://www.dbgroup.unimo.it/tesi
13. Charalambis, E., Bergamaschi, S., Jagadish, H.V.: Bayesian networks: optimization of the human-computer interaction process in a Big Data scenario (2014). https://www.dbgroup.unimo.it/tesi
14. Gámez, J.A., Mateo, J.L., Puerta, J.M.: Learning Bayesian networks by hill climbing: efficient methods based on progressive restriction of the neighborhood. Data Min. Knowl. Disc. **22**(1–2), 106–148 (2011)
15. Meng, X., Bradley, J., Yavuz, B., Sparks, E., Venkataraman, S., Liu, D., Freeman, J., Tsai, D.B.. Amde, M., Owen, S., et al.: MLlib: Machine Learning in Apache Spark (2015). arXiv:1505.06807
16. Scott, S.L., Blocker, A.W., Bonassi, F.V., Chipman, H.A., George, E.I., McCulloch, R.E.: Bayes and Big Data: the consensus Monte Carlo algorithm. Int. J. Manage. Sci. Eng. Manage. **11**, 77–88 (2016)
17. Fang Q., Yue K., Fu X., Wu H., Liu W.: A MapReduce-based method for learning Bayesian network from massive data. In: Ishikawa Y., Li J., Wang W., Zhang R., Zhang W. (eds.) Web Technologies and Applications. APWeb 2013. Lecture Notes in Computer Science, vol. 7808, pp. 697–708. Springer, Berlin (2013)

## References

# Big Data Meet Pharmaceutical Industry: An Application on Social Media Data

**Caterina Liberati and Paolo Mariani**

**Abstract** Big Data are hard to capture, store, search, share, analyze, and visualize. Without any doubts, Big Data represent the new frontier of data analysis, although their manipulation is far to be realized by standard computing machines. In this paper, we present a strategy to process and extract knowledge from Facebook data, in order to address marketing actions of a pharmaceutical company. The case study relies on a large Italians sample, interested in wellness and health care. The results of the study are very sturdy and can be easily replicated in different contexts.

**Keywords** Big Data · Dimensions reduction · Knowledge extraction · Healthcare sector

## 1 Introduction

With term Big Data (BD), we usually refer to a collection of data sets so large and complex that standard analytical tools are unable to process. The Big Data are hard to capture, store, search, share, analyze, and visualize. Their presence is increasing in recent years, due to the massive amount of machine data generated every day from mobile devices, tracking systems, radio-frequency identification, sensor networks, social networks, Internet searches, automated record keeping, video archives, e-commerce, etc. It is neither just a matter of scale, BD have changed radically how we think about getting knowledge from the data (reframing completely

The original version of this chapter was revised: New figure replacement. The erratum to this chapter is available at https://doi.org/10.1007/978-3-319-55708-3_27

C. Liberati (✉) · P. Mariani
DEMS, Universitá degli Studi di Milano-Bicocca,
p.zza Dell'Ateneo Nuovo 1, 20126 Milan, Italy
e-mail: caterina.liberati@unimib.it

P. Mariani
e-mail: paolo.mariani@unimib.it



F. Mola et al. (eds.), *Classification, (Big) Data Analysis and Statistical Learning*, Studies in Classification, Data Analysis, and Knowledge Organization, https://doi.org/10.1007/978-3-319-55708-3_3

**Table 1** Data versus Big Data: the 7 V

| V | Data | Big Data |
|---|---|---|
| Volume | Megabyte MB - $10^6$ | Zettabyte ZB - $10^{21}$ |
| Velocity | Static | Real time |
| Variety | Structured and rarely integrated from different sources | Structured and unstructured. Not integrated. Collected from different sources |
| Variability | Low | High |
| Veracity | High | Low |
| Visualization | Simple | Complex |
| Value | High | Not verified |

the processes of research), how we should engage with information, and the nature and the categorization of reality [1]. The market sees Big Data as pure opportunity. Companies are interested in digital technology and Big Data because they can spot new market trends, they can provide insights in a variety of fields, from Information Technology (IT) to medicine to law enforcement and everything in between and beyond.

Without any doubts, Big Data represent the new frontier of data analysis, although it still persists a substantial confusion between having a lot of data available and working in a contest of BD. In order to clarify such misunderstanding, we can take into account seven[1] characteristics [5] that allow us to highlight differences and to uncover peculiarities of each type of information (Table 1).

In the light of the social changing that is occurring due to BD, new rules and principles will need to cover six broad areas: privacy, security, retention, processing, ownership, and the integrity of information [6]. In particular, referring to health-care system, a recent trend is observed: the sector is moving toward an evidence-based medicine, where not only doctors but computers are involved in diagnosis and patients' monitoring [7]. In this context, the digitalization of records would make it much easier to spot and monitor health trends and evaluate the effectiveness of different treatments. On the other hand, large-scale efforts to computerize health records tend to run into bureaucratic, technical, and ethical problems [8]. As in other fields, also in the health care, huge and complex volumes of data are present. Everything ranging between health care to patients' well-being can be defined as Big Data in health sector: diagnostic imaging, laboratory results, pharmacy purchases, insurance information, electronic health records but also the messages of social media, blogs, status updates on Facebook and other platforms, and web pages. Main difficulties, related to their usage, lie in both the inability to employ traditional tools to synthesize and interpret results, and in the speed of processing and managing different sources

[1]In several discussions and articles, Big Data are attributed to have three Vs: Volume, Velocity, and Variety [2, 3]. Such definition has been updated by [4] that added two more characteristics to the data: Value and Veracity.

of information [9]. In other words, we have all the problems related to analyze Big Data. In this regard, the pharmaceutical industry is beginning to explore BD contest, although these efforts are still in their early stages [10].

The aim of this study is propose a method to explore and synthesize Big Data in order to get insights about potential customers. Inspired by a report produced by Cubeyou on Facebook data, referred to the pharmaceutical sector [11], we analyzed micro-data, applying statistical technique for sparse matrices reducing. The rest of the paper is organized as follows: Sect. 2 provides a description of the research design and the collected data. Section 3 illustrates the modeling employed and the main results. Finally, Sect. 4 proposes some concluding remarks and possible further developments.

## 2 The Goal and the Data

The data of the research come from Cubeyou, a social media company which helps businesses in structuring marketing activities and addressing informed marketing decisions in the areas of media and content. The data were collected at the end of 2014 among Italian users of Facebook interested in wellness and health. Thus, the selection of the instances has been realized keeping those users have visited pages as Bristol-Myers Squibb, Amgen, Boehringer Ingelheim, Schering-Plough, Baxter International, Takeda Pharmaceutical, AIFA, just to cite a few. We collected all the possible interactions among people and brands, products, and services (i.e., shares, likes, tweets, pins, posts, etc.).

Of course, such huge amount of information could not be handled and processed with standard computing engineering. Therefore, the raw data were stored on a cloud platform with 20 servers active on Amazon Web Services (AWS) infrastructure. More than 5 Terabyte (distributed) databases were gathered and updated daily via Hadoop2. Hadoop is an open-source software designed to handle extremely high volumes of data in any structure. It was composed of two elements:

1. the Hadoop Distributed File System (HDFS), which supports data in structured relational form, in unstructured form, and in any form in between
2. the MapReduce programming paradigm for managing applications on multiple distributed servers with the focus on supporting redundancy, distributed architectures, and parallel processing

In particular, MapReduce is a development framework created by Google to perform parallel computations and it consists of two phases: the first one decomposes a problem down into subproblems and delegates them to other machines (nodes) that can do the same thing recursively, producing outputs organized by set of key of values. The second one attributes results of the lower nodes to the upper ones. This calculates a partial result (reduction) that combines all the values for the same key, and so on.

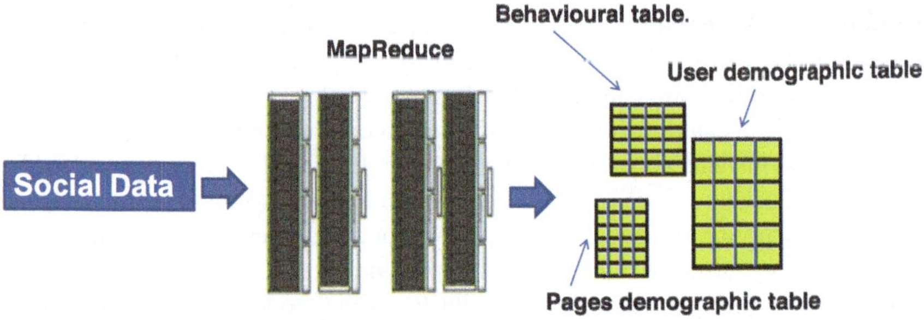


**Fig. 1** MapReduce process flow

In our case, the synthesis process stored information into three main tables: the *Behavioral Table* a huge dataset of more than 8 billions rows that contained each user by Facebook page interactions, the *User Demographic Table* that collects unstructured data about users profiles, and the *Pages Demographic Table* that stores unstructured data about Facebook pages (Fig. 1).

In each table, records were extracted with queries based on users' keys and behavior. In addition to what we have already illustrated, each instance has a psychographic[2] category obtained via a classification algorithm. The classifier,[3] at first developed by a collaboration between Cubeyou and Cambridge University, has been trained on Facebook likes provided by over 58,000 volunteers, together with their detailed demographic records and synthesis of several personality tests [12]. Therefore, the unstructured data as personality traits, psycho-graphical data, user location, hobbies, and interests have been transformed in meaningful customer information. The application of logistic regression has provided the label for each record of our sample, so subjects have been distinguished among: Pet Lovers, Outdoor Enthusiast, Techies, Car Lovers, Book Lovers, Social Activist, Gamers, Movie Lovers, Politically Active, Sport Lovers, Fashion Lovers, Music Lovers, Travel Lovers, Public Figures Followers, Food Lovers, Home Decorators, Beauty and Wellness Aware, Business People, and Housekeepers.

The ambitious goal of our work is to propose a strategy to explore and synthesize the same raw dataset but using only the Facebook information, in order to aid pharmaceutical businesses to make their strategic decisions in terms of communication or activities and targets. The matrix, provided by the IT infrastructure, had 5607 rows (Facebook users) and 159 dummies columns (19 psychographic profiles and 140 topics). The columns collected the users' likes on several themes, relevant for comparing characteristics and preferences from a socio-behavioral perspective: TV

---

[2]Psychographic data classify people based on demographic characteristics but using their interests, attitudes, habits, values, and opinions, not only on their objective, in order to better understand what drives them to purchase and engage with the company. The basic assumption under this practice lays on the fact that products and/or brands purchased by an individual express part of his/her imaginary and personal preferences.

[3]The classifier employed in the analysis is exclusive Cubeyou's ownership.

channels, magazines, celebrities, online resources, and so on. For sake of brevity, we report in Sect. 3 only those evidences relative to preferences to public figures, but more insights could be produced.

We work under the assumption that each user can set more than one like per page. The application aims to explore perceptions of public figures for a possible media campaign.

## 3 Methodology and Results

Before considering techniques or models, we have to face the question: "Are we working with Big Data?" The answer is "Yes" if we consider the origin of the database, "No" if we consider the present format. Of course, the initial size of the table would have not allowed to run any statistical analysis and so MapReduce process, which collapsed the BD in a manageable matrix, was necessary. The second question to address is "How do we elaborate this table?" One option was Multiple Correspondence Analysis, but the sparsity of the data would have effected the results. Such issue is really common in social media data that collect a deluge of data on people (where they are, who they are, how they live, and how they buy) but that are unable to provide an overview on market segmentation. Therefore, we run a preprocess step, in order to reshape the table. The matrix dimensions have been reduced, generating a contingency table of likes. The squeezing process provided a matrix of 140 topics and 19 psychographics profiles.[4] As already illustrated, the frequency cells of the contingency table are not sums of binary preferences, each individual has had the option to set more than one like per page.

We employed a Principal Component Analysis [13]. We run preliminary checks as Kaiser-Meyer-Olkin test (0.949) and the sphericity Bartlett's test (8911.841 df=171, $p$ value=0.000) that confirmed the suitability of the technique implementation to our case. We retained only those factors with eigenvalue greater or equal to 1, in order to generate a principal plane which explains the 93.65% of the total inertia. Visual inspection of the variables' coordinates onto the first two principal components (Table 2) addressed the naming process of the two axes. The first one has been interpreted as *Hedonism* (47.97% of explained inertia), due to its high correlations with outdoor enthusiast, car lovers, gamers, movie lovers, sports lovers, music lovers, beauty, and wellness aware profiles. The second axis, called *Commitment*, (45.68% of explained inertia) reports pet lovers, techies, book lovers, activist company, politically active, business people, and home decorators.

[4]The squeezing process changed the subjects of the analysis: we do not refer to the micro-records anymore but to large groups of people that fit into known categories. Such process does not cause a loss of information because the objective of the analysis is related to orientate media choices into a marketing communication plan. In such context, it not necessary to profile each single user but it is useful get characteristic groups with alternative desires needs and attitudes.

**Table 2** Factorial coordinates of the psychological profiles onto the principal plane

| | Factor 1 | Factor 2 |
|---|---|---|
| Pet lovers | 0.507 | 0.832 |
| Outdoor enthusiast | 0.788 | 0.542 |
| Techies | 0.477 | 0.863 |
| Car lovers | 0.910 | 0.385 |
| Book lovers | 0.508 | 0.820 |
| Social activist | 0.543 | 0.834 |
| Gamers | 0.876 | 0.323 |
| Movie lovers | 0.810 | 0.573 |
| Politically active | 0.297 | 0.928 |
| Sport lovers | 0.879 | 0.433 |
| Fashion lovers | 0.689 | 0.618 |
| Music lovers | 0.823 | 0.555 |
| Travel lovers | 0.740 | 0.648 |
| Public figures followers | 0.660 | 0.736 |
| Food lovers | 0.717 | 0.687 |
| Home decorators | 0.756 | 0.631 |
| Beauty and wellness aware | 0.737 | 0.533 |
| Business people | 0.497 | 0.848 |
| Housekeepers | 0.577 | 0.653 |

Other profiles as housekeepers, public figures followers, fashion lovers, food lovers, and travel lovers are transversal respect to the two factors. Therefore, the graphical representation of the Italian public figures onto the Hedonism versus Commitment plane expresses (indirectly) the perceptions of FB users, interested in drugs and health, about such celebrities (Fig. 2). In the first quadrant of the principal plane,[5] characterized by positive Hedonism and Commitment values, we found Marco Travaglio, Fiorello, Beppe Grillo, and Luciana Littizzetto. The second one, with Hedonism negative and Commitment positive values, shows Gino Strada, Papa Francesco, and Massimo Gramellini. The third, with Hedonism and Commitment both negative, collects Sonia Peronaci e Giulio Golia. Finally, in the fourth quadrant, where Hedonism positive and Commitment negative, there are Marco Bazzoni, Belen Rodriguez, Paolo Bitta, Alessia Marcuzzi, and Alessandro Borghese.

[5]Authors elaboration on Cubeyou data—November 2014.

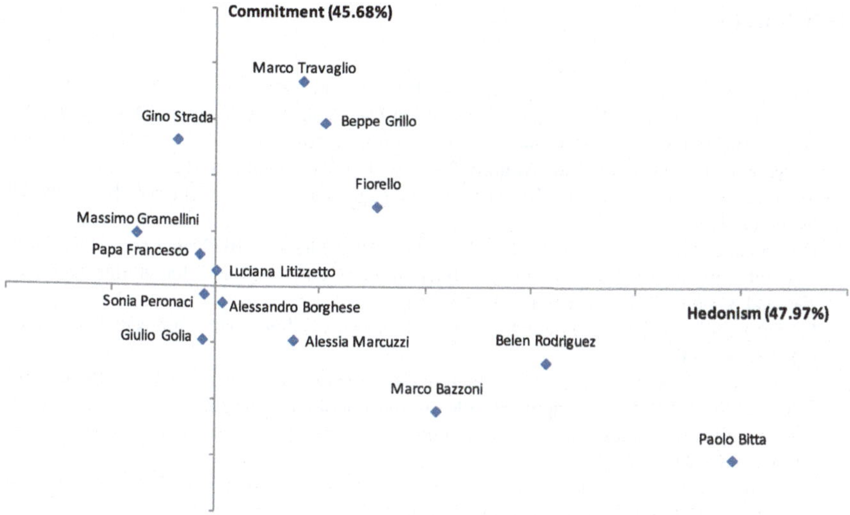


**Fig. 2** Public figures projected onto Hedonism versus Commitment plane

# 4 Further Developments

Social media data offer a set of users' information that can be declared, if they are spontaneous or directly self-reported by the person (as gender, age, job place, etc.) or gathered, if they are derived from the interactions with other contents (as businesses pages, users, etc.) [14]. Social media data are Big Data when the seven V are respected. In our case, the modeling of the data was performed after preprocess that has legitimated the usage of the explorative factorial analysis. Results of our study uncover two main findings: (1) Big Data can meet pharmaceutical industry's marketing needs; (2) The sample of Italian people interested in pharmaceutical environment is very small. As expected, in such sample, most of the users show peculiar attention to wellness (interpreted as general concept, not related to drugs or pathologies). We limited the analysis to the public figures but the plenty of the information available from the survey could be modeled in further works.

The results are very interesting, especially if we consider the context: pharmaceutical industries are attempting to measure adherence on therapies and R&D area. To this purpose, marketing departments of such companies are focusing their attention to Big Data [15]. The objective of the companies in the short term is to extract knowledge from this large amount of data, in order to implement a series of activities that might improve the quality of health care [16]. Even the use of a testimonial could be crucial for a prevention campaign or to increase compliance with a therapy.

In the light of all these considerations, the usage of BD could produce public benefits as reducing costs to support clinical decisions and a more efficient management of public health.

## References


1. Boyd, D., Crawford, K.: Critical questions for Big Data. Inf. Commun. Soc. **15**, 662–679 (2012)
2. Douglas, L.: 3D Data Management: Controlling Data Volume, Velocity and Variety. Technical report, META Group. http://blogs.gartner.com/doug-laney/files/2012/01/ad949-3D-Data-Management-Controlling-Data-Volume-Velocity-and-Variety.pdf (2001)
3. McAfee, A., Brynjolfsson, E.: Big Data: the management revolution. Harvard Bus. Rev. **10**, 59–68 (2012)
4. Demchenko, Y., Grosso, P., de Laat, C., Membrey, P.: Addressing Big Data issues in Scientific Data Infrastructure. Proceedings of 2013 International Conference on Collaboration Technologies and Systems (CTS), IEEE, pp. 48–55 (2013)
5. McNulty-Holmes, E.: Understating Big Data: the seven V's. http://dataconomy.com/seven-vs-big-data/ (2014)
6. The Economist: New rules for Big Data. http://www.economist.com/node/15557487 (2010)
7. Falotico, R., Liberati, C., Zappa, P.: Identifying oncological patient information needs to improve e-health communication: a preliminary text-mining analysis. Qual. Reliab. Engng. Int. **31**, 1115–1126 (2015)
8. The Economist: The data deluge. Businesses, governments and society are only starting to tap its vast potential. http://www.economist.com/node/15579717 (2010)
9. Lazer, D., Kennedy, R.: What We Can Learn From the Epic Failure of Google Flu Trends. Wired. http://www.wired.com/2015/10/can-learn-epic-failure-google-flu-trends (2015)
10. Santoro, E.: Web 2.0 e Medicina: come social network, podcast, wiki e blog trasformano la comunicazione, l'assistenza e la formazione in sanitá. Il pensiero scientifico Editore, Milano (2009)
11. Cubeyou: How pharmaceutical market customers benchmark against average Italian people, Industry Report, pp. 1–36 (2014)
12. Kosinski, M., Stillwell, D., Graepel, T.: Private traits and attributes are predictable from digital records of human behaviour. Proc. US Natl. Acad. Sci. **110**(15), 5802–5805 (2013)
13. Bolasco, S.: Analisi Multidimensional dei dati. Carocci, Roma (2010)
14. Moubarak, G., Guiot, A., Benhamou, Y., Hariri, S.: Relationship and its impact on the doctor-patient Facebook activity of residents and fellows. J. Med. Ethics. **37**, 101–104 (2010)
15. Greene, J.A., Kesselheim, A.S.: Pharmaceutical marketing and the new social media. N. Engl. J. Med. **363**, 2087–2089 (2010)
16. Mariani, P., Mussini, M.: L'integrazione di dati amministrativi e campionari per arricchire i sistemi informativi di marketing nel farmaceutico: evidenze empiriche legate al record linkage. Micro Macro Mark. **2**, 295–318 (2014)

# Electre Tri Machine Learning Approach to the Record Linkage

**Valentina Minnetti and Renato De Leone**

**Abstract** This paper proposes, for the first time in the literature, the application of the Electre Tri method for solving the record linkage matching. Results of the preliminary stage show that, by using the Electre Tri method, high accuracy can be achieved and more than 99% of the *matches* and *nonmatches* are correctly identified by the procedure.

**Keywords** Multiple criteria classification · Linked data · Linear programming

## 1 Linked Data: The Record Linkage

The aim of this paper is to propose the multiple criteria Electre Tri method for solving the Record Linkage (RL) matching [1, 2]. RL is the methodology of bringing together corresponding records from two or more files or finding duplicates within files [3]. The methods solving the RL matching were conceived to tackle the absence or the unreliability of the identifier variables and the presence of errors, missing data in the matching variables [4].

A small practical example is now presented, in order to explain the issue more effectively. Suppose two datasets of persons T and S (reported in Tables 1 and 2, respectively), whose variables are Name, Address, and Age, have to be linked. In this example, the identifiers are not known and the records contain errors in some digits, but not missing values. Referring to the variables Name in both datasets, the values "John A Smith" (of T) and "J H Smith" (of S) should probably refer to the

V. Minnetti (✉)
Department of Statistic Science, Faculty of Information Engineering, Informatics and Statistics, Sapienza University of Rome, Rome, Italy
e-mail: valentina.minnetti@uniroma1.it

R. De Leone
School of Science and Technology, Section of Mathematics, University of Camerino, Camerino, Italy
e-mail: renato.deleone@unicam.it



F. Mola et al. (eds.), *Classification, (Big) Data Analysis and Statistical Learning*, Studies in Classification, Data Analysis, and Knowledge Organization,
https://doi.org/10.1007/978-3-319-55708-3_4

**Table 1** Data in the dataset T

| Name | Address | Age |
|---|---|---|
| John A Smith | 16 Main Street | 16 |
| Javier Martinez | 49 E Applecross Road | 33 |
| Gillian Jones | 645 Reading Aev | 22 |

**Table 2** Data in the dataset S

| Name | Address | Age |
|---|---|---|
| J H Smith | 16 Main St | 17 |
| Haveir Marteenez | 49 Aplecross Road | 36 |
| Jilliam Brown | 123 Norcross Blvd | 43 |

same person (i.e., *match*), as well as "Javier Martinez" (of T) and "Haveir Marteenez" (of S). In this case, the probabilistic method is used. Such a method tackles the RL matching problem with the statistical test theory, by verifying if each pair of units in the matching table is generated either by the distribution of the *matches* or by the distribution of *nonmatches* [4]. The statistical test is the following likelihood ratio:

$$R_{ts} = \frac{Pr((t,s) \in \Gamma \mid M)}{Pr((t,s) \in \Gamma \mid U)}, \tag{1}$$

where $\Gamma = T \times S$ is the Cartesian product of the tables $T$ and $S$, $M$ is the set of *matches* and $U$ is the set of *nonmatches*. Since for a subset of pair $(t, s)$ the classification can be uncertain, the space $\Gamma$ can be partitioned into three subsets: $M$, $U$, and $P$, where $P$ contains the pairs with no sufficient information to take a decision.

Fixing the errors of I type $\alpha$ and $\beta$ that can be tolerated (*matches* that are classified as *nonmatches* and *nonmatches* classified as *matches*), the classification of each pair $(t, s)$ is made as follows (decision rule) [4]:

- if $R_{ts} > T_\beta$, then the pair $(t, s)$ is a *match*;
- if $R_{ts} < T_\alpha$, then the pair $(t, s)$ is a *nonmatch*;
- if $T_\beta \leq R_{ts} \leq T_\alpha$, then the pair $(t, s)$ is an *uncertain match*;

where $T_\beta$ and $T_\alpha$ are two thresholds (upper and lower, respectively) such that $T_\beta > T_\alpha$. In practice, these thresholds are fixed arbitrarily [5], even if they depend on the errors $\alpha$ and $\beta$.

From the decision rule, the RL can be viewed as a classification problem, with three classes, fixed a priori [2]. The classification methods for solving the RL matching, currently used or under consideration by statistical worldwide agencies, are the following supervised learning techniques: classification tree [6, 7], Support Vector

Machines [8, 9], and Neural Networks [10]. In this work, another supervised learning technique with multiple criteria approach is considered. It is described in the following section.

## 2 The Multiple Criteria Electre Tri Method: A Brief Description

In this section, the main features of the Multiple Criteria Decision Aiding (MCDA) and the Electre Tri method are introduced.

In MCDA, a finite set of $n$ objects (alternatives, actions and projects) $A = \{a_1, a_2, \ldots, a_n\}$ is evaluated by means of a finite set of $m$ criteria $G = \{g_1, g_2, \ldots, g_m\}$. A criterion is the real-valued function $g_j : A \rightarrow \Re$, whereby $g_j(a_k)$ indicates the performance of the alternative $a_k$ on the criterion $g_j$. In the literature, there are many types of criteria, such as the real-criterion, the pseudo-criterion, the quasi-criterion, and interval-criterion. Moreover, a criterion $g_j$ can be either of gain or cost type. In the first case, the Decision-Maker (DM) prefers high values of $g_j$, while in the second case, the DM prefers low values of $g_j$. The comparison of any pair of alternatives $a_i$ and $a_k$, by means of a binary relation, can be performed by comparing the values $g_j(a_i)$ and $g_j(a_k)$ [11].

Electre Tri is a MCDA method that solves the ordinal sorting problem, that is, a classification problem in which the categories are ordered in a strict preference relation. The classification of each alternative to a category is the result of the comparisons between the alternative and the profiles, in order to achieve an outranking relation. The profiles are "fictitious alternatives" separating the categories. Formally, if all criteria are of the gain type, given $p$ categories $C_1, C_2, \ldots, C_p$, ordered such that $C_1 \prec C_2 \prec \cdots \prec C_p$, the profile $b_h$ is the upper limit of the category $C_h$ and the lower limit of category $C_{h+1}$. In this case, $C_1$ and $C_p$ are the worst and the best categories, respectively. For example, comparing the alternative $a_i$ with the profile $b_h$, the outranking relation $a_i S b_h$ validates the assertion "$a_i$ *outranks* $b_h$" whose meaning is "$a_i$ is at least as good as $b_h$". However, the outranking relations do not imply automatically the assignments. Two possible assignment procedures are proposed: the pessimistic one and the optimistic one. For example, with the pessimistic procedure, the alternative $a_i$ is assigned to the category $C_{h+1}$ iff the relation $a_i S b_h$, for $h = p-1, p-2, \ldots, 1$, is validated.

In the context of the Electre Tri method, the validations of all the outranking relations are made by means of the computations of four indices [11, 12]:

1. the partial concordance indices on each criterion;
2. the global concordance index on all the criteria;
3. the partial discordance indices on each criterion;
4. the credibility index on all the criteria.

The constructions of the indices 1 and 3 have to take into account that Electre Tri is a method based on a pseudo-criterion. The pseudo-criterion works as follows:

- if $|g_j(a_i) - g_j(b_h)| \le g_j(q_h)$, then $a_i I_j b_h$ and $b_h I_j a_i$;
- if $g_j(q_h) < |g_j(a_i) - g_j(b_h)| < g_j(p_h)$, then $a_i Q_j b_h$ and $b_h Q_j a_i$;
- if $|g_j(a_i) - g_j(b_h)| \ge g_j(p_h)$, then $a_i P_j b_h$ and $b_h P_j a_i$;

where $I_j$ represents the indifference binary relation ($\sim$), $Q_j$ is the weak preference binary relation ($\succeq$), $P_j$ is the strict preference binary relation ($\succ$) on the criterion $g_j$, $g_j(q_h)$ is the indifference and $g_j(p_h)$ is the preference thresholds, to be fixed.

Suppose all the criteria are of gain type and the pessimistic procedure is adopted. In order to find the outranking relations $a_i S b_h$, the comparisons are made on the left side of the profile $b_h$. With these assumptions, the partial concordance index on the criterion $g_j$ indicates how much such criterion is in accordance with the statement "$a_i$ *outranks* $b_h$". The structure of the pseudo-criterion implies that this index takes three values, one in each interval. Thus, the profiles, the preference, and the indifference thresholds values must be fixed in advance.

All the partial concordance indices are then aggregated into the global concordance index. It is the weighted arithmetic mean of the partial concordance indices, weighted by the weights $w_1, w_2, \ldots, w_m$. They represent the importance coefficients of each component of the mean value. The partial discordance index on the criterion $g_j$ indicates how much such criterion is opposed to the statement "$a_i$ *outranks* $b_h$". Also, this index takes three values, in the three intervals established by the preference and the veto thresholds. The veto threshold represents the smallest difference $|g_j(a_i) - g_j(b_h)|$ over all criteria, incompatible with the assertion "$a_i$ *outranks* $b_h$". For the computation of the partial discordance index, the profiles, the preference, and the veto thresholds values must be fixed in advance.

The credibility index synthesizes the concordance and the discordance indices. It corresponds to the global concordance index weakened by the veto effects. So, if the veto thresholds do not enter into the classification model, the credibility index is equal to the global concordance index. The cutting level $\lambda$ is the minimum credibility index value that permits to state the outranking relation, as follows:

$$if \quad \sigma(a_i, b_h) \ge \lambda \quad then \quad a_i S b_h.$$

Briefly, the parameters of the Electre Tri to be estimated are the profiles, the thresholds (preference, indifference, and veto), the weights, and the cutting level. They can be elicited directly; in this way, it is hard to have a clear global understanding of the implications of these values in terms of the output [12]. If they are elicited indirectly, in order to have a control on the parameters' values, an estimation procedure in two or more phases is recommended. In the literature, such a procedure exists, that is, the two-phase procedure estimating all the parameters of the Electre Tri [2, 13, 14]. The first phase is devoted to the profiles and the thresholds estimations; the second one to the weights and the cutting level estimations. In this procedure, thresholds are considered as linear functions of the profiles, while cutting level is a nonlinear function of the weights [2, 13, 14]. Since the weights are estimated in the second phase, depending on the results obtained in the first phase of the procedure, so the only parameters to be inferred are the profiles.

Given a training set, composed of assignments examples, i.e., $a_k \to C_h$, the profiles are inferred by solving the following Linear Programming (LP) problem:

$$\min z = \sum_{j=1}^{m} \sum_{a_k \to C_h} \theta_j(a_k) \quad \text{subject to}$$

$$\begin{array}{lll}
\theta_j(a_k) \geq g_j(a_k) - g_j(b_h) & \forall j = 1, \ldots, m, \forall a_k \to C_{h,h \neq p} & (1) \\
\theta_j(a_k) \geq g_j(b_{h-1}) - g_j(a_k) & \forall j = 1, \ldots, m, \forall a_k \to C_{h,h \neq 1} & (2) \\
g_j(b_h) \geq g_j(b_{h-1}) + \varepsilon & \forall j = 1, \ldots, m, \forall h = 2, \ldots, p-1 & (3) \\
\theta_j(a_k) \geq 0 & \forall j = 1, \ldots, m, \forall a_k \to C_h & (4)
\end{array}$$

where $\varepsilon$ is a small positive value, arbitrarily fixed [2, 13, 14]. The problem consists in the minimization of the sum of all the classification errors $\theta_j(a_k)$ on the alternatives belonging to the training set. Constraints (1) and (2) define the error $\theta_j(a_k)$ when the alternative $a_k$ on the criterion $g_j$ lies outside the category declared in the training set. In the case of large size problems, it is useful to compute the measure of the range of the profile $b_h$, before the first phase of the procedure, as follows [2]:

$$\varphi_j(b_h) = \min_{a_s \to C_{h+1}} \{g_j(a_s)\} - \max_{a_v \to C_h} \{g_j(a_v)\} \quad \forall h = 1, \ldots, p-1, \forall j = 1, \ldots, m$$

If $\varphi_j(b_h) < 0$ for at least one criterion, then the profiles must be determined by solving the LP problem, defined above [2].

## 3 Application to Real Data: A Preliminary Stage

The input data, used for the proposed application, were taken from the Simulated List of people from the American Census (in the SecondString file for approximate string matching techniques). Two datasets A and B are provided, containing 449 and 392 units, respectively. The true links are 327 (true matches).

The variables for each dataset are the following: DS (*labels of the datasets: A and B*); IDENTIFIER; SURNAME; NAME; LASTCODE (*middle name initial*); NUMCODE (*address street number*); and STREET (*address street name*).

Aim of the application is to determine a classifier which assigns each record of the matching dataset $A \times B$ to one of the three categories $C_1$ (*nonmatches*), $C_2$ (*uncertain matches*), and $C_3$ (*matches*), by using the Electre Tri method. This method was chosen because it is based on a pseudo-criterion. This type of criterion is suitable for the case where data are affected by errors due to uncertainty and/or imprecisions. The parameters of the Electre Tri were estimated by means of the two-phase procedure, described above. With this procedure, a training set is required. The test set method, recommended when big databases are analyzed, was used [1, 2].

The application is still being developed, due to its complexity. Here, results from the first preliminary stage are reported; the priority was to test if the Electre Tri method was suitable for finding good classifiers. As a consequence, all the choices were made in order to simplify the analysis. For example, since in both datasets there are records with missing values, in the preliminary stage, all these records were deleted. But how to treat them is the aim of further investigations. So, suppose that the subset of the matching dataset $A' \times B' \subset A \times B$ was the input data, where A$'$ contains 282 and B$'$ 298 units. Thus, the number of the true links decreases to 243.

The variables DS, in both datasets, were not considered in the application. The variables IDENTIFIER were assumed to be not available. The strategy was to use the variables IDENTIFIER in order to find the three categories and so the training set and the test set [1, 2]. The set of all the records in the matching table was partitioned into four subsets, as follows:

- $S_1$ contains the records with $IDENTIFIER_A \neq IDENTIFIER_B$ and $STREET_A \neq STREET_B$ (79763 units);
- $S_2$ contains the records with $IDENTIFIER_A = IDENTIFIER_B$ and $STREET_A \neq STREET_B$ (56 units);
- $S_3$ contains the records with $IDENTIFIER_A = IDENTIFIER_B$ and $STREET_A = STREET_B$ (187 units); and
- $S_4$ contains the records with $IDENTIFIER_A \neq IDENTIFIER_B$ and $STREET_A = STREET_B$ (4030 units).

With this partition, the categories are defined as follows: $S_1 \cup S_4 = C_1$ (*nonmatches*), $S_2 = C_2$ (*potential matches*), and $S_3 = C_3$ (*matches*).

Since the cardinality of $S_2$ is the smallest one, then only 56 records must be chosen in each subset such that the sample is "in equilibrium" [1, 2]. In this way, the training set contained 168 units and test set 83868 units. With four subsets, it is possible to create two different training sets from $S_1$, $S_2$, $S_3$ (used for finding classifiers M1) and from $S_2$, $S_3$, $S_4$ (used for finding classifiers M2).

Another important choice is the distance measure. Such a measure permits to transform the initial matching dataset into the dataset of alternatives-criteria. In the literature, there are several distance measures [5, 15].

In the preliminary stage of the application, a new distance measure was proposed: the Generalized Equality Measure (GEM) [1, 2]. It is defined as follows:

$$g_j(a_n) = \frac{\sum_{i=1}^{min\{length[v_j(t_k)], length[v_j(s_m)]\}} I_j^i(t_k, s_m)}{max\{length[v_j(t_k)], length[v_j(s_m)]\}},$$

where $I_j^i(t_k, s_m) = \begin{cases} 1 & if\ v_j^i(t_k) = v_j^i(s_m) \\ 0 & if\ v_j^i(t_k) \neq v_j^i(s_m) \end{cases}$

where $v_j^i(t_k)$ indicates the element of the unit $t_k$ in the position $i$ on the variable $j$. With the use of GEM, all criteria are always of the gain type and binary relationships between the categories are always of the form $C_3 \succ C_2 \succ C_1$.

Before the first phase of the two-phase procedure, the values of the measures of the range of the two profiles $b_1$ and $b_2$ indicated that the LP problem was solved. Once $\varepsilon = 0.01$ was fixed, the LP problem with 1125 constraints and 850 variables was solved in less than one second by computing up to 1000 iterations. In the first phase, after having estimated the profiles, the thresholds were considered as suggested in [2, 13, 14]. The reported results do not consider the veto thresholds. Of course, if the veto thresholds are considered, other results are obtained [1].

The second phase provides a system of nonlinear inequalities, only for the alternatives not automatically assigned, of the following form:

$$\begin{cases} \sigma(a_k, b_1) \geq \lambda, \forall a_k \rightarrow \{C_2, C_3\} \\ \sigma(a_k, b_2) \geq \lambda, \forall a_k \rightarrow C_3 \end{cases}$$

For example, if $\underline{w} = [1 \;\; 1 \;\; 1 \;\; 1 \;\; 1]'$, in order to find classifiers that well classify all the *matches*, then necessarily $\lambda \leq 0.60$. With those parameters, the performances of the classifiers M1 and M2 are reported in Table 3, which shows the total number of misclassified alternatives on the training set, on the test set and on all data:

**Table 3** Total number of misclassified alternatives found by the classifiers M1 and M2

| Classifiers | M1 | | | M2 | | |
|---|---|---|---|---|---|---|
| **Training set** | $C_1$ | $C_2$ | $C_3$ | $C_1$ | $C_2$ | $C_3$ |
| $C_1(56)$ | – | 1 | 0 | – | 2 | 2 |
| $C_2(56)$ | 0 | – | 47 | 0 | – | 47 |
| $C_3(56)$ | 0 | 1 | – | 0 | 1 | – |
| Tot (168) | 0 | 2 | 47 | 0 | 3 | 49 |
| Tot misclass | 49 | | | 52 | | |
| Accuracy (%) | 70.83 | | | 69.05 | | |
| **Test set** | $C_1$ | $C_2$ | $C_3$ | $C_1$ | $C_2$ | $C_3$ |
| $C_1(83737)$ | – | 455 | 304 | – | 128 | 302 |
| $C_3(131)$ | 0 | 3 | – | 0 | 3 | – |
| Tot (83868) | 0 | 458 | 304 | 0 | 131 | 302 |
| Tot misclass | 762 | | | 433 | | |
| Accuracy (%) | 99.09 | | | 99.48 | | |
| **All data** | $C_1$ | $C_2$ | $C_3$ | $C_1$ | $C_2$ | $C_3$ |
| $C_1(83793)$ | – | 456 | 304 | – | 130 | 304 |
| $C_2(56)$ | 0 | – | 47 | 0 | – | 47 |
| $C_3(187)$ | 0 | 4 | – | 0 | 4 | – |
| Tot (84036) | 0 | 460 | 351 | 0 | 134 | 351 |
| Tot misclass | 811 | | | 485 | | |
| Accuracy (%) | 99.03 | | | 99.42 | | |

$\underline{w} = [1 \;\; 1 \;\; 1 \;\; 1 \;\; 1]'$ and $\lambda = 0.50$

( ) is the cardinality of the corresponding set on the left side. Both classifiers M1 and M2 classify almost all the *matches* to the right category ($C_3$). Only four alternatives are classified as *potential matches*. The value 456 in M1 is the total number of misclassified alternatives that are assigned to the *potential matches* category ($C_2$) instead of the *nonmatches* category ($C_1$). This value decreases to 130 when M2 is the classifier. The other values of misclassified alternatives almost coincide for both classifiers. This decrease shows that the choice of the alternatives in $S_4$ instead of in $S_1$ implies a remarkable improvement. The accuracy index for each set (training set, test set, and all data) defines the performance of the classifiers in the referred set. For example, the value 99.48% measures how many alternatives are well classified on the cardinality of the test set (83868). If $\lambda$ increases in the test set, the number of misclassified alternatives in the *nonmatches* category decreases, the number of misclassified alternatives in the *matches* category increases and the accuracy index increases [1]. In fact, as reported in [1], if $\lambda = 0.70$, then in the test set the accuracy index is 99.81 and 99.85% for M1 and M2, respectively; if $\lambda = 0.85$, then in the test set the accuracy index is 99.89% for both classifiers. The convergence to the value 99.89% is slower for M2 than for M1, because for $\lambda = 0.50$ the accuracy index is higher for M2 than that for M1.

The choice of the most interesting classification model is left to the DM, depending on his/her preferences, taking into account that the *matches* category $C_3$ is the preferred one.

## 4 Conclusions

For the first time, the Electre Tri method has been proposed for solving the RL. The parameters of the Electre Tri are estimated by means of the two-phase procedure, requiring a training set. Only the results of a preliminary stage are reported. At least by judging from the preliminary stage, the multiple criteria classification method Electre Tri, applied for solving the RL, has found good classifiers. With these encouraging results, the proposed application answers to one of many challenges in applying the supervised machine learning to RL matching [16]. Moreover, the first preliminary stage confirms that the RL matching is highly sensitive to the quality of preprocessing [17]. Surely, the distance measure and the training set play the most important roles, contributing to obtain good classifiers. Even if the GEM measure, for its simple formulation, could not lead to interesting results, good results were obtained with the fundamental contribution of the Electre Tri, due to its mathematical structure and to the strategy used in the construction of the training set. In other words, these good results are systematic if and only if the assumptions are combined together in the construction of the classification model. These assumptions are the use of GEM, the strategy in the construction of the training set, the use of the Electre Tri method, and the use of the two-phase procedure.

In the future experiments, other distance measures as well as different schemes in the training set, taking into account missing data, will be tested.

The importance of this application is due to the increasing development of the use of administrative data (also combined with survey data). In this context, an important problem is that of finding matching pairs of records from heterogeneous databases, while maintaining privacy of the databases parties. In view of this aim, secure computation of distance metrics is important for secure RL [15].

## References

1. Minnetti, V.: A new distance measure for solving Record Linkage with Electre Tri. Technical Report (February, 2015)
2. Minnetti, V.: On the parameters of the Electre Tri method: a proposal of a new two phases procedure. Ph.D. thesis on Operational Research, Sapienza University of Rome (2015)
3. Winkler, W.E.: The state of Record Linkage and current research problems. U.S. Bureau of the Census, available at: https://www.census.gov/srd/papers/pdf/rr99-04.pdf (1999)
4. Fellegi, I.P., Sunter, A.B.: A theory for record linkage. J. Am. Stat. Assoc. **64**, 1183–1210 (1969)
5. Cibella, N., Fortini, M., Scannapieco, M., Tosco, L., Tuoto, T., Valentino, E.: RELAIS User's Guide, Version 2.2, available at www.istat.it/it/files/2011/03/Relais2.2UserGuide.pdf (2010)
6. Cohen, W.W.: The WHIRL approach to data integration. IEEE Intell. Syst. **13**(3), 20–24 (1998)
7. Elfeky, M., Elmagarmid, A.: TAILOR: a record linkage toolbox. In: Proceedings of the 18th International Conference on Data Engineering (ICDE'02), pp. 17–28 (2002)
8. Bilenko, M., Mooney, R.: Adaptive duplicate detection using learnable string similarity. In: Proceedings of the 9th ACM SIGKDD International Conference on Knowledge Discovery and Data Mining (KDD), pp. 39–48 (2003)
9. Christen, P: Automatic training example selection for scalable unsupervised record linkage. In: Washio, T., Suzuki, E., Ting, K.M., Inokuchi, A. (eds.), Advances in Knowledge Discovery and Data Mining: 12th Pacific-Asia Conference, PAKDD, pp. 511–518 (2008)
10. Wilson, D.R.: Beyond probabilistic record linkage: using neural network and complex features to improve genealogical record linkage. In: Proceedings of IEEE International Joint Conference on Neural Networks, pp. 9–14 (2011)
11. Mousseau, V., Slowinski, R., Zielniewicz, P.: ELECTRE TRI 2.0, a methodological guide and user's manual. Document du LAMSADE no111, Universit Paris-Dauphine (1969)
12. Mousseau, V., Slowinski, R.: Inferring an ELECTRE TRI model from assignment examples. J. Glob. Optim. **12**, 157–174 (1998)
13. De Leone, R., Minnetti, V.: New approach to estimate the parameters of Electre Tri model in the ordinal sorting problem. In: Proceedings of AIRO 2011—Operational Research in Transportation and Logistics, p. 69 (2011)
14. De Leone, R., Minnetti, V.: The estimation of the parameters in multi-criteria classification problem: the case of the electre tri method. In: Vicari, D., Okada, A., Ragozini, G., Weihs, C. (eds.) Analysis and Modeling of Complex Data in Behavioral and Social Sciences, pp. 93–101. Springer, Berlin (2014)
15. Cohen, W.W., Ravikumar, P., Fienberg, S.E: A Secure protocol for computing string distance metrics. In: Proceedings of IEEE International Conference on Data Mining (ICDM), pp. 40–46 (2004)
16. Chu, K., Poirier, C.: Machine Learning Documentation Initiative. https://www.unece.org/fileadmin/DAM/stats/documents/ece/ces/ge.50/2015/Topic3_Canada_paper.pdf (2015)
17. Winkler, W.E.: Matching and record linkage. WIREs Comput. Stat. **6**, 313–325 (2014)

# Part II
# Social Networks

# Finite Sample Behavior of MLE in Network Autocorrelation Models

**Michele La Rocca, Giovanni C. Porzio, Maria Prosperina Vitale and Patrick Doreian**

**Abstract** This work evaluates the finite sample behavior of ML estimators in network autocorrelation models, a class of auto-regressive models studying the network effect on a variable of interest. Through an extensive simulation study, we examine the conditions under which these estimators are normally distributed in the case of finite samples. The ML estimators of the autocorrelation parameter have a negative bias and a strongly asymmetric sampling distribution, especially for high values of the network effect size and the network density. In contrast, the estimator of the intercept is positively biased but with an asymmetric sampling distribution. Estimators of the other regression parameters are unbiased, with heavy tails in presence of non-normal errors. This occurs not only in randomly generated networks but also in well-established network structures.

**Keywords** Network effect model · Density · Network topology · Non-normal distribution

---

M. La Rocca (✉) · M. P. Vitale
Department of Economics and Statistics, University of Salerno, Fisciano, Italy
e-mail: larocca@unisa.it

M. P. Vitale
e-mail: mvitale@unisa.it

G. C. Porzio
Department of Economics and Law, University of Cassino
and Southern Lazio, Cassino, Italy
e-mail: porzio@unicas.it

P. Doreian
Faculty of Social Sciences, University of Ljubljana, Ljubljana, Slovenia
e-mail: pitpat@pitt.edu

P. Doreian
Department of Sociology, University of Pittsburgh, Pittsburgh, USA



F. Mola et al. (eds.), *Classification, (Big) Data Analysis and Statistical Learning*,
Studies in Classification, Data Analysis, and Knowledge Organization,
https://doi.org/10.1007/978-3-319-55708-3_5

## 1 Introduction

Network autocorrelation models (NAMs) [1–3] deal with the presence of individual units embedded within social structures. They represent a class of auto-regressive models used to study the effect of a network on an outcome variable of interest when the data points are interdependent. Specifically, we can refer to the "social influence" (or contagion) mechanism in which the social relations among individuals provide a foundation for predicting actor behaviors given the behaviors of other actors in the network in which they are embedded [4].

Among the models proposed in the literature to address social influence on individual behavior, NAMs propose an approach dealing with, simultaneously, network effects and individual attributes. Despite the clear advantages over other conventional approaches, it is known that in these models the estimated autocorrelation parameter has a finite sample negative bias, the amount of which is positively related with the network density [5, 6].

Our contribution aims at describing the *whole* finite sample distribution of the Maximum Likelihood Estimators (MLE) of the autocorrelation and regression parameters. Through an extensive simulation study, we investigate the conditions under which MLEs are normally distributed in the finite sample case. The finite sample distributions are evaluated with respect to the network density and topology, the distribution of error terms, and the strength of the autocorrelation parameter (i.e., the network effect size).

We focus on three research questions:

- *What is the whole sampling distribution of the network effect estimator?*
- *What are the finite sample distributions of the regression coefficient estimators?*
- *What are the consequences of the errors not being normally distributed?*

The remaining of the paper is organized as follows. Section 2 presents a brief review on NAMs. The Monte Carlo simulation study used to deal with the aforementioned research questions is described in Sect. 3. Section 4 reports the main results, while Sect. 5 concludes with some final remarks.

## 2 A Brief Review of Network Autocorrelation Models

Two types of network autocorrelation models are available within the literature [1]: the network effects model and the network disturbances model. In the first case, interdependencies between actors are modeled through the inclusion of an autocorrelation parameter in the dependent term while in the second case, interdependencies are included in the disturbance term. Here, the focus is on the network effects model which allows individual outcome to be directly associated with neighbors' levels of outcome by including the network effect as a weight matrix [7].

More formally, let $\mathbf{y}$ be a $(n \times 1)$ vector of values of a dependent (endogenous) variable for $n$ individuals making up a network, let $\mathbf{X}$ represent the $(n \times p)$ matrix of values for the $n$ individuals on $p$ covariates (including a unit vector for the intercept term), and let $\mathbf{W}$ be the $(n \times n)$ network weight matrix whose elements, $w_{ij}$, measure the influence actor $j$ has on actor $i$. The network effects model is defined as

$$\mathbf{y} = \rho \mathbf{W} \mathbf{y} + \mathbf{X}\beta + \varepsilon$$

where $\rho$ is the network autocorrelation parameter referred as the strength of the social influence mechanism in a network, $\beta$ is a $(p \times 1)$ vector of regression parameters, and the error terms $\varepsilon$ are assumed to be independently normally distributed with zero means and equal variances, $\varepsilon \sim N(0, \sigma^2 \mathbf{I})$.

This class of models represents a popular tool for conducting social network analysis. First adopted to describe social influence mechanism [4], it has been recently applied in many social science fields (see e.g., [8–11]) and it has been extended to the study of multiple networks [12, 13] and to the presence of two-mode networks [14].

From a methodological point of view, recent contributions focused on the bias of the MLE of the network autocorrelation parameter $\rho$. They discovered a systematic negative bias, whose magnitude increases with the network density [5]. In addition, it has been found that this bias does not depend on network size, numbers of exogenous variables in the model, and whether the network weight matrix $\mathbf{W}$ was normalized or reported in raw form. The bias also does not depend on the presence of well-established network structures (e.g., scale-free and small-world configurations), although it is especially pronounced at extremely low-density levels in the star network [6].

Recently, rather than looking for more conditions in which network autocorrelation parameter is underestimated, Wang et al. [15] investigated the likelihood of identifying a statistically significant network effect. They found that first Type error rates are substantially controlled by the Wald procedure. In addition, they had that the statistical power of the test is a non-linear increasing function of $\rho$ and of the network size, while it is not particularly related to the density and to the network structure. Faber et al. [16] showed that the average degree of a random network impacts the power of tests.

However, as highlighted within the introduction, knowing the full sampling distributions of MLE estimates is needed. This is the focus of our study. Its design and results are described in the next sections.

## 3 Simulation Design

An extensive Monte Carlo study (5000 MC replications) was used to assess the whole finite sampling distribution of MLEs. To accomplish this, the following conditions have been varied: (i) the network density ($\Delta$), (ii) the network autocorrelation parameter ($\rho$), (iii) the network topology ($\mathbf{W}$), and (iv) the error distributions ($\varepsilon$).

Two covariates were considered, and data were generated according to the following network autocorrelation model:

$$y_i = \rho W y_i + \beta_0 + \beta_1 x_{1i} + \beta_2 x_{2i} + \varepsilon_i, \qquad i = 1, 2 \ldots, n.$$

Corresponding to the simulation design used by [5, 6], the elements in the simulation scheme were set as follows. Covariates were randomly generated according to a standard normal distribution; without loss of generality, all the $\beta$'s were set equal to the same constant value ($\beta_0 = \beta_1 = \beta_2 = 2$); only positive values of $\rho$ were considered ($0.00 \leq \rho \leq 0.90$), accounting for low-to-high network effect size. The error was independently randomly generated with constant variance ($Var[\varepsilon_i] = 1$) according to three different schemes. First, as it is usually done in such studies, errors were standard normally generated. Second, to consider the effect of asymmetric errors, a standard lognormal was considered. Finally, to consider a completely nonstandard distribution, the error term was derived by generating data from an equal weight mixture of distributions of a (centered) chi-square with one degree of freedom and a student's t with four degrees of freedom.

We consider a sample size of 50 nodes. The network weight matrix **W** was row normalized and randomly generated at each run. The network density $\Delta$ took values $0.05 \leq \Delta \leq 0.80$. Beyond the Erdos–Renyi random graphs (E-Rs) adopted as baseline model, two other kinds of topologies were considered to assess the evidence of nonrandom behaviors in the formation of network ties. Specifically, the scale-free [17] and the small-world [18] network configurations were taken into account so that the effect of well-established network structure can be evaluated, as reported in [6]. In the first case, preferential attachment defines the tie formation mechanism. This mechanism accounts for the tendency to be linked with the best connected nodes (i.e., nodes with the highest degree). Hence, a scale-free structure emerges when nodes' degree distribution follows a power law distribution, and a "star" network structure appears (i.e., one node is linked to all the others and no other connections are present among the remaining nodes). The small-world configuration presents instead a high node connectivity with low average distance among regions of the network. More specifically, the concurrent presence of dense local clustering (measured by clustering coefficient) with short network distances (measured by the average path length) is observed.

## 4 Results

We examine the observed sampling distributions of $(r - \rho)$, $(b_0 - \beta_0)$, $(b_k - \beta_k)$, where $r$, $b_0$ and $b_k$ are the ML estimators of $\rho$, $\beta_0$, and $\beta_k$, respectively, (k = 1, 2). First, somewhat to our surprise, all three of the examined network structures provided substantially the same results. One implication is that the operation of network effects in NAMs does not rest on the global structure of the network. For this reason, only results for the E-R random graph are discussed further.

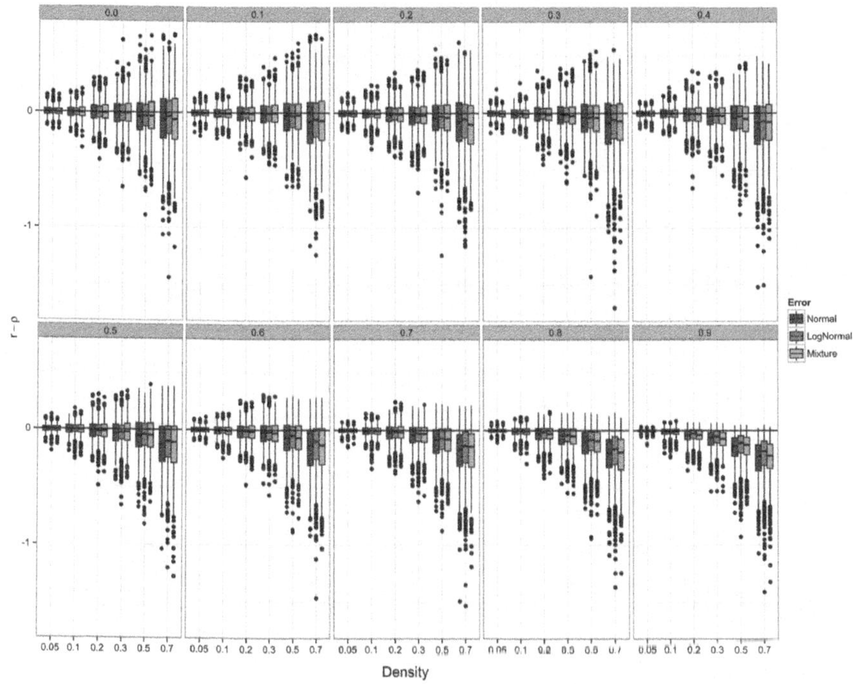


**Fig. 1** Boxplots of the sampling distribution $r - \rho$ (vertical axis) for E-R random graphs. The density values are reported on the horizontal axis. Results are reported for different values of $\rho$

Parallel boxplots are reported in each picture illustrating the ML finite distribution for different levels of network effects size, densities, and error distributions.

The sampling distributions obtained for $(r - \rho)$ are shown in Fig. 1. Each frame in the figure corresponds to simulation results obtained for a fixed value of the population parameter $\rho$ ($\rho$=0, 0.1, 0.2,..., 0.9). Boxplots show results with respect to five level of density (horizontal axes, $\Delta$=0.05,..., 0.8), and three error distributions (normal, lognormal, mixture, in this order).

As expected, sampling distributions are negatively biased, a result increasing with $\rho$ and $\Delta$. Normality does not seem to hold: for low values of $\rho$ and $\Delta$, heavy tails appear. For higher values of both parameters, distributions are quite strongly asymmetric. Finally, it seems that differences in the error distributions have a minor effect on the resulting estimator distributions.

Adopting the same graphical structure, results obtained for the sampling distributions of the regression coefficient estimators are reported in Figs. 2 and 3 [for $(b_0 - \beta_0)$ and $(b_1 - \beta_1)$, respectively]. Results for the sampling distributions of $(b_2 - \beta_2)$ are not reported as they are the same to those observed for $(b_1 - \beta_1)$.

As for the regression coefficient estimators, some different behaviors arise. The intercept estimator distributions mirrors the distribution of the autocorrelation parameter estimator: it is positively biased, with such a bias increasing with $\rho$ and $\Delta$,

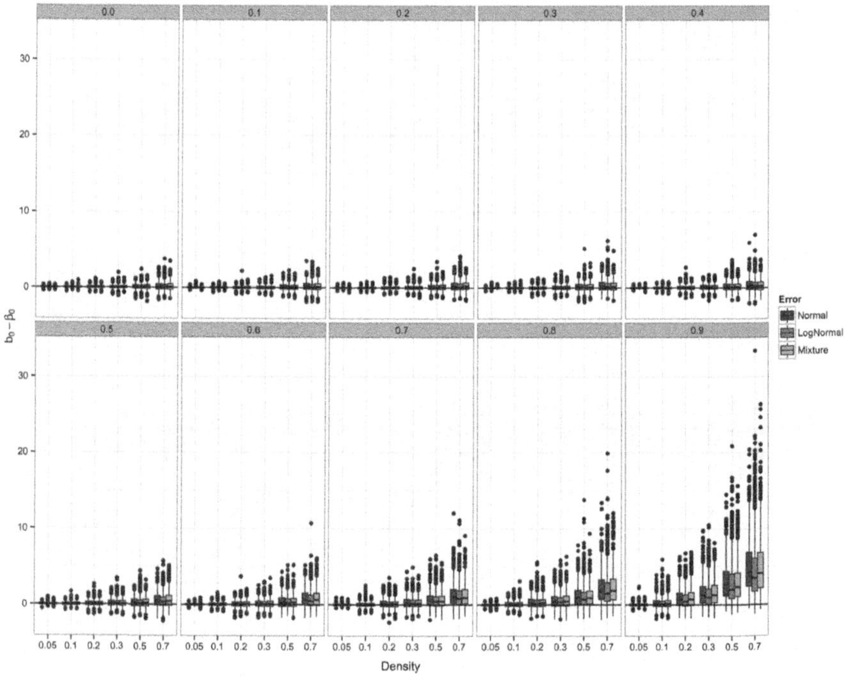


**Fig. 2** Boxplots of the sampling distribution of $b_0 - \beta_0$ (vertical axis) for E-R random graphs. The density values are reported on the horizontal axis. Results are reported for different values of $\rho$

with analogous effects in terms of asymmetries and heavy tails. On the other hand, results suggest that the estimators of the other regression coefficients are unbiased, with heavy tails in the presence of non-normal errors.

Overall, it seems that where non-normality of the estimator distributions arises as a consequence of a certain degree of autocorrelation and density, this effect overwhelms the effect due to non-normality of the errors. However, where distributions are unbiased and normal, non-normality of the errors plays a more substantial role. To conclude, ML estimators of the autocorrelation parameter and of the intercept are not normally distributed in case of small sample size, even in presence of normally distributed errors. Furthermore, the network density has some effect on the variability of the estimators. On the other hand, it seems that other features of the network topologies, in the main, have little effects.

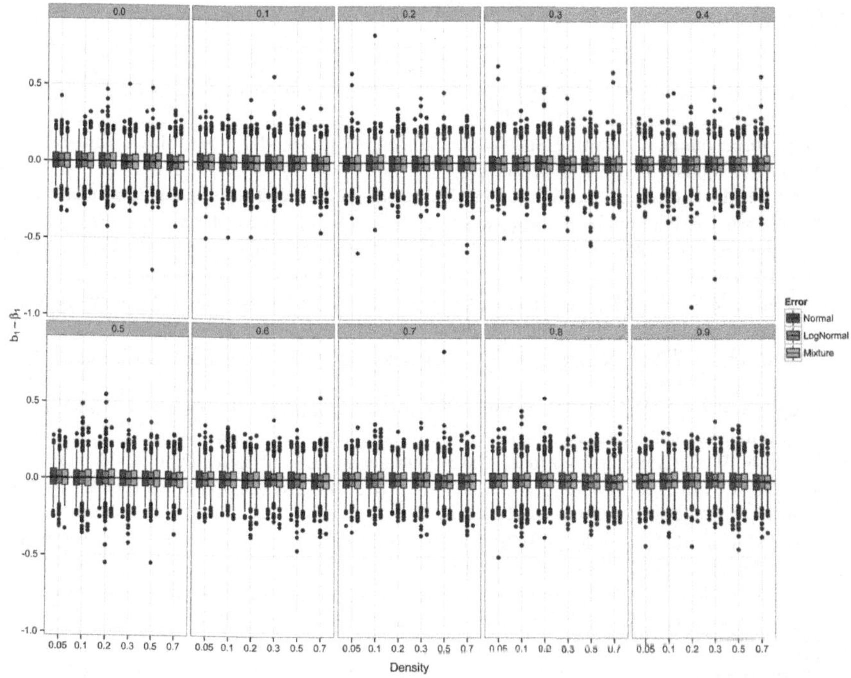


**Fig. 3** Boxplots of the sampling distribution of $b_1 - \beta_1$ (vertical axis) for E-R random graphs. The density values are reported on the horizontal axis. Results are reported for different values of $\rho$

## 5 Discussion and Conclusions

The present contribution has shown that the ML estimator of $\rho$ in NAMs not only contains a systematic negative bias, as expected, but also that its distribution is typically non-normal and asymmetric.

According to our results, for high values of the autocorrelation parameter $\rho$ and network density $\Delta$, the sampling distribution of the autocorrelation parameter is negatively biased and quite strongly asymmetric. On the other hand, the sampling distribution of regression coefficients is positively biased and asymmetric for the estimator of the intercept, and unbiased and with heavy tails in the presence of non-normal errors for the other regression coefficients. This occurs not only in randomly generated networks but also in well-established network structures as well.

Furthermore, the non-normality and the asymmetry are not confined strictly to networks with high density. At least in small-world networks, these features also exist for very low levels of density. This suggests to study the performances of other related estimation tools, such as the finite sample confidence interval behaviour. The authors intend to report on that issue elsewhere.

## References


1. Doreian, P.: Linear models with spatially distributed data: spatial disturbances or spatial effects? Sociol. Methods Res. **9**, 29–60 (1980)
2. Doreian, P., Teuter, K., Wang, C.H.: Network autocorrelation models: some Monte Carlo results. Sociol. Methods Res. **13**, 155–200 (1984)
3. Dow, M.M., Burton, M.L., White, D.R.: Network autocorrelation: a simulation study of a foundational problem in regression and survey research. Soc. Netw. **4**, 169–200 (1982)
4. Marsden, P.V., Friedkin, N.E.: Network studies of social influence. Sociol. Methods Res. **22**, 127–151 (1993)
5. Mizruchi, M.S., Neuman, E.J.: The effect of density on the level of bias in the network autocorrelation model. Soc. Netw. **30**, 190–200 (2008)
6. Neuman, E.J., Mizruchi, M.S.: Structure and bias in the network autocorrelation model. Soc. Netw. **32**, 290–300 (2010)
7. Leenders, R.T.A.: Modeling social influence through network autocorrelation: constructing the weight matrix. Soc. Netw. **24**, 21–47 (2002)
8. De Nooy, W.: Communication in natural resource management: agreement between and disagreement within stakeholder groups. Ecol. Soc. **18**(2), 44 (2013). https://doi.org/10.5751/ES-05648-180244. (Available via DIALOG)
9. Dow, M.M., Eff, E.A.: Global, regional, and local network autocorrelation in the standard cross-cultural sample. Cross Cult. Res. **42**, 148–171 (2008)
10. Franzese Jr., R.J., Hays, J.C., Kachi, A., Alvarez, R.M., Freeman, J.R., Jackson, J.E.: Modeling history dependence in network-behavior coevolution. Polit. Anal. **20**, 175–190 (2012)
11. Vitale, M.P., Porzio, G.C., Doreian, P.: Examining the effect of social influence on student performance through network autocorrelation models. J. Appl. Stat. **43**, 115–127 (2016)
12. Dow, M.M.: Galton's problem as multiple network autocorrelation effects cultural trait transmission and ecological constraint. Cross Cult. Res. **41**, 336–363 (2007)
13. Zhang, B., Thomas, A.C., Doreian, P., Krackhardt, D., Krishnan, R.: Contrasting multiple social network autocorrelations for binary outcomes with applications to technology adoption. ACM T. Man. Inf. Syst. **3**, 1–21 (2013)
14. Fujimoto, K., Chou, C.P., Valente, T.W.: The network autocorrelation model using two-mode data: affiliation exposure and potential bias in the autocorrelation parameter. Soc. Netw. **33**, 231–243 (2011)
15. Wang, W., Neuman, E.J., Newman, D.A.: Statistical power of the social network autocorrelation model. Soc. Netw. **38**, 88–99 (2014)
16. Farber, S., Páez, A., Volz, E.: Topology and dependency tests in spatial and network autoregressive models. Geogr. Anal. **41**, 158–180 (2009)
17. Albert, R., Barabási, A.-L.: Statistical mechanics of complex networks. Rev. Mod. Phys. **74**, 47–97 (2002)
18. Watts, D.J., Strogatz, S.H.: Collective dynamics of 'small-world' networks. Nature **393**, 440–442 (1998)

# Network Analysis Methods for Classification of Roles

**Simona Gozzo and Venera Tomaselli**

**Abstract** This paper concerns *network analysis* (NA) methods employed to classify role profiles based on the measurement and the specification of linking structures among pupils in infant school. With this aim, we measure sociometric status by the direct observation of preschool children behaviours. The data have been collected by a longitudinal observational study. We follow the children in seven waves for 2 months. In this study, we measure relational skills of subjects applying three procedures: The regular equivalence is used to detect similar positions within the network, the lambda sets to observe the ability to be intermediary and the cliques to assess the propensity to belong to a group. Concerning each analytic dimension, the results show an increase of relational competence and an association with ego-network measures.

**Keywords** *Network analysis* · Sociometric status · *Structural equivalence* · *Lambda sets* · *Cliques*

## 1 Introduction

Several studies have shown that peer relationships have significant effects on language development, learning abilities and empathetic capabilities [9, 14].

On the basis of this assumption, our study analyses the capability of preschool children (3–5 years old) to relate with peers in building behaviour patterns consistent over time so as to rise social roles.

We monitored the relational behaviour of 42 pupils aged from 3 to 5 years, obtaining a panel survey repeated for seven waves, during 2 months (October and December)

S. Gozzo · V. Tomaselli (✉)
Department of Political and Social Sciences, University of Catania,
Vitt. Emanuele II 8, 95131 Catania, Italy
e-mail: tomavene@unict.it

S. Gozzo
e-mail: simonagozzo@yahoo.it



F. Mola et al. (eds.), *Classification, (Big) Data Analysis and Statistical Learning*,
Studies in Classification, Data Analysis, and Knowledge Organization,
https://doi.org/10.1007/978-3-319-55708-3_6

by directly observing the pupils' behaviours in their school setting, with the aim to measure sociometric status in order to classify social roles.

The concept of role suggests a structure of relationships among the subjects in a network. In a NA approach, the notion of social role depends conceptually, theoretically and formally on the specific relationships that link the set of actors and their positions across the network.

## 2 Theoretical Framework

In our hypothesis, children in preschool age (3–5 years old) have relational skills sufficiently developed to allow the analysis of sociometric status [4–6, 12, 13].

The literature focuses on two scores: acceptance and rejection, defined by the number of the most liked and least liked nominations. Combining these two kinds of nominations, social preference and social impact indicators are obtained [4]. Thus, the following sociometric status are identified:

- *popular*: altruistic children, who comply with social norms, have high social impact and preference (i.e. many contacts and many positive relationships)
- *rejected*: excluded by other children because of their aggressiveness or self-isolation (they have little or no contact and negative relationships)
- *controversial* (not observed in the present study): aggressive, in spite of they define themselves as popular, avoiding peer rejection and isolation (high social impact and high index of aggressiveness)
- *average*: 60% of interviewed pupils falls into this category. They are less friendly than the popular but more sociable than the rejected
- *neglected*: usually with only a good relationship, often repeated, with the same subject (best friend).

## 3 Materials and Methods

The main aim of our study is to classify social roles by means of NA. We employ three NA tools: *regural equivalence*, *lambda sets* and *cliques*.

Specifically, in order to identify the sociometric status of the nodes in a network, we apply the *regular equivalence* (REGE algorithm), the most suitable for directed and weighted matrices. According to White and Reitz [16], a graph $G$ is an ordered pair $(P, R)$, where $P$ is a finite set of points (here, children) and $R$ are ties on $P$, that is a subset of the ordered pairs of points in $PXP$. As a consequence, the *regular equivalence* for single-relation networks is defined as

if $G = (P, R)$ and $\equiv$ is an equivalence relation on $P$,
then $\equiv$ is a *regular equivalence* if and only if for all $a, b, c \in P$,
$a \equiv b$ involves that

($i$) $aRc$ implies there exists $d \in P$ such that $bRd$ and $d \equiv c$

(*ii*) *cRa* implies there exists $d \in P$ such that *dRb* and $d \equiv c$.

REGE is an iterative algorithm. Within each iteration, a search is implemented to optimize a matching function between nodes or vertices *i* and *j*. As a consequence, for each *k* in *i*'s neighbourhood, the algorithm searches for an *m* in *j*'s neighbourhood of similar value.

A measure of similar values is based upon the absolute difference among sizes of ties. The measure is then weighted by means of the degree of equivalence between *k* and *m* at the previous iteration. The matching is thus optimized. This is summed for all members of *i*'s neighbourhood overall relations and normalized to provide the current iteration measure of equivalence between *i* and *j*. The procedure is repeated for all the pairs of vertices for a fixed number of iterations.

The result of this iterative procedure is a similarity matrix, which provides a measure of *regular equivalence*. This matrix is then submitted to a clustering routine using the single hierarchical linkage. We thereby obtain a *dendrogram* of the regular similarity measure with the level at which any pair of actors are aggregated [16].

According to the literature about sociometric status [4–6, 12, 13], the similarity matrix and REGE algorithm are applied to cluster children identifying social roles.

Based on the assumption that the members of a cluster have greater edge connectivity with other members than with non-members, we measure the ability to be mediators in a relational context by the *lambda set*. Given a graph *G* (*V*, *E*), where *V* is a finite set of points (or actors) and *E* are ties on *V*, a *lambda set S* represents the edge connectivity of two points *a* and *b* in the graph. It is a subset of *V*, such that for all

$$a, b, c \subset S, d \in (V - S) => \lambda(a, b) > \lambda(c, d). \tag{1}$$

Computing a *lambda set*, we can obtain a matrix of partitions in which a value of *K* in an *i*-column and in a *j*-row indicates that the node *j* is in the *k*-partition and the other members of the partition form a *lambda set* with minimum edge connectivity. This information allows to derive a permutation of the nodes used in a particular hierarchical clustering of the nodes in a network, properly corresponding to a *lambda set* [2].

Finally, a *clique* is the maximum complete sub-graph composed of three or more nodes. It is, therefore, a maximal subset of nodes in which each node is directly connected with each other [11]. A *clique S* of a graph *G (V, E)* is a maximal subset (*alpha*) of *V*, such that for all

$$s \in S, \alpha(s, S) = |S| - 1. \tag{2}$$

By means of *cliques*, we identify the structure of the cohesive partitions within the networks, allowing to observe the propensity of each node to be part of a group. Indeed, the same node or set of nodes may belong to different, also overlapped, cliques with a score as high as the number of occurrences in a *clique* [3]. In this study, the *cliques* are used to describe relational abilities aimed to identify the propensity of each pupil to be part of a group.

## 4 Results: The Classification of Social Roles

Our research defines the roles on a monthly basis.

Starting from October, pupils are less connected with each other, since this is the first month of school. The socialization is beginning. As a consequence, we observe a lack of repeated and important ties among children. In December, the analysis shows more ties with an increasing reciprocity and a denser net.

Furthermore, in the same month, the data structure permits to use the criterion of *regular equivalence*, applying the REGE algorithm [10], the most suitable according to the nature of the data.

In literature, three types of nodes are defined [1]: *sinks*, with only incoming ties; *repeaters*, with both incoming and outgoing ties and *sources*, with only outgoing ties. In our analysis, children are classified according to four profiles:

1. *isolated* and *invisible*, grouped in December (similar to *rejected* and *neglected*)
2. *sinks*, more sociable (similar to *popular*)
3. *repeaters*, with an almost equal number of ties to and from the nodes (similar to *average*)
4. *sources*, the 'reference points', more often contacted by others than contacting others (similar to *popular*).

In comparison with the cognitive approach based on interview, our method allows to better specify *popular* sociometric status, splitted into *sinks* and *sources*.

The second method to analyse the relational competence is the *lambda set*. This procedure employs a partition criterion, based on the greater or lesser capability of the subjects to be 'intermediaries', to classify pupils in three groups for each month, with the following relational traits:

- *connectors*, supporting the whole structure of the network
- *bridges*, intermediaries between subgroups or individuals
- *marginals*, playing no strategic role as intermediaries.

From October to December, many nodes move from a group to another, increasing the number of bridges in December and decreasing connectors. Only *marginal* nodes tend to remain in the same group: children, unable to be intermediaries in October, will rarely become in December.

In order to select groups of children, we employ *cliques*. This procedure identifies subjects with a good ability to build cohesive subgroups. In October, the number of cliques is negligible: only 17 nodes show relational competence. It is, therefore, impossible to identify subgroups in this month. In December, by contrast, the structure becomes more complex (Fig. 1).

The number and profile of the groups in the network change while the relational competence increases, showing three groups:

1. *embedded*, children located in the majority of overlapping *cliques*, thanks to their ability to develop closed subgroups within the net

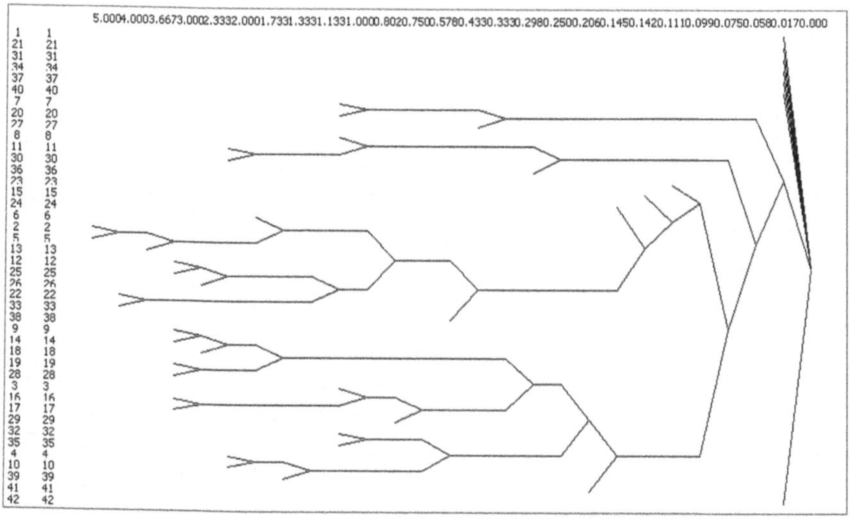


**Fig. 1** The dendrogram of the cliques of December

2. *integrated*, children with the ability to be part of simple, small groups, consisting of at least a triad
3. *marginals*, with weak capacity to belong to a group.

As further step in our analysis, we classify the relational dynamics by the criterion of role. The question is: can we detect a relationship between individual characteristics and social roles? In order to check this assessment, we employ the Kruskal–Wallis test introducing three blocks of variables, see Table 1:

1. structural attributes
2. sociability indicators
3. ego-network measures.

The latter block is referred to links to and from each node but not to the whole network.

We check the significance of each attribute referred to every cluster, identified by means of *lambda sets* since it is the unique instrument that shows relevant results both for October and December. First results showed that in December, there is an increasing number of *bridges*, while *marginals* pupils decrease much more than *connectors* do.

By means of the Kruskal–Wallis test, see Table 2, we notice the underline dynamics. Compared with October, in December the aggressiveness is not significant and a larger number of positive contacts rather than negative are not associated with the ability to act as intermediary. However, both positive and negative relationships are equally associated with the broker ability.

**Table 1** Variables employed for Kruskal–Wallis test

| Label | Description |
|---|---|
| *GenderM, GenderF* | Dichotomous variables: gender |
| *Indicators of sociability* | |
| *Pos rel* | Number of positive relationships |
| *Neg rel* | Number of negative relationships |
| *INDEX social impact* | Total number of contacts per node |
| *INDEX social preference* | Difference between positive and negative relationships |
| *INDEX social aggressiveness* | Difference between negative and positive relationships |
| *$\alpha$-parameter* | Parameter detected by the $p^1$ model estimating the sociability as out-degree of each node |
| *$\beta$-parameter* | Parameter detected by the $p^1$ model, estimating the attractiveness as in-degree of each node |
| *Measures of ego-network* | |
| *EffSize* | Overall impact of the ego in the network[a] |
| *Efficiency* | Impact of the ego on the network for each new tie[b] |
| *Size* | Number of nodes (subjects) related to ego |
| *Ties* | Number of ties of the ego |
| *Ordered pairs* | Number of possible ties of the ego |
| *Density* | Percentage of ties observed over the total number of possible ties |
| *nBroker* | Normalized brokerage capacity, i.e. ability of mediation of the ego[c] |
| *nEgoBe* | Normalized *Betweenness*: capacity of each ego to connect subjects/groups each other compared with the total capacity to connect subjects/groups in the network (*Betweenness* of the whole network) |

[a]Difference between the number of nodes with whom ego has ties and average number of links that each of nodes has with other nodes
[b]Relationship between *EffSize* and the size of the ego-network
[c]This ability is measured by the number of pairs of neighbours not directly connected to each other divided by the total number of possible pairs, in order to remove the effect of the network size

Summing-up, from October to December, pupils have quite similar ego-network values but we notice an increase for every significant parameter and, in particular, for sociability and attractiveness ($\alpha$ and $\beta$). This is probably due to the increasing number of *connectors*. Namely, *connectors* are the most attractive ($\beta$-parameter) and *bridges* are the most sociable ($\alpha$-parameter).

**Table 2** Kruskal–Wallis test: role of 'broker' by lambda sets

| Roles by lambda sets | October | December |
|---|---|---|
| | ($\chi^2$) | ($\chi^2$) |
| *GenderM* | 0.72 | 8.91** |
| *Pos rel* | 18.45*** | 12.35** |
| *Neg rel* | 2.82 | 9.01 |
| *INDEX social impact* | 11.17** | 15.32*** |
| *INDEX social preference* | 6.27* | 0.771 |
| *INDEX social aggressiveness* | 6.27* | 0.771 |
| *EffSize* | 33.57*** | 34.96*** |
| *Efficiency* | 15.02** | 0 |
| *Size* | 34.15*** | 36.68*** |
| *Ties* | 24.04*** | 26.09*** |
| *Ordered pairs* | 34.15*** | 36.68*** |
| *Density* | 10.02** | 1.18 |
| *nBroker* | 9.45** | 1.47 |
| *nEgoBe* | 0.264 | 6.43* |
| *α-parameter* | 16.35*** | 21.39*** |
| *β-parameter* | 10.21** | 23.53*** |
| *Lambda: connectors October* | | 6.29* |

Only significant values are shown: *significant for $\alpha = 0.05$; **significant for $\alpha = 0.01$; ***significant for $\alpha = 0.001$

## 5 Conclusions

In order to detect the presence of social roles and explore the structure of the relationships among children, we carried out the research employing the method of direct observation. This choice is due to the considerable gap between verbal and relational competence.

In our findings, we detect a coherent evolution along the time, in terms of relational structure and role building of pupils observed. The network of December shows an increasing relevance of 'brokers' and 'reference points' or 'popular' children and only few isolated children as well as high levels of interaction. An increasing specialization in the individual roles even among children aged 3–5 is implied.

Finally, in our study, the *regular equivalence* procedure provides results quite similar to those already known in literature. Instead, *lambda sets* and *cliques* enrich the assessment of the traditional sociometric status detecting further suggestive relational dynamics.

The research could be developed by different approaches [8, 15], also investigating the effect of the external auxiliary variables on the relational data structures [7].

## References


1. Borgatti, S.P., Everett, M.G.: The class of all regular equivalences: algebraic structure and computation. Soc. Netw. **11**, 65–88 (1989)
2. Borgatti, S.P., Everett, M.G., Shirey, P.R.: LS sets, lambda sets and other cohesive subsets. Soc. Netw. **12**, 337–357 (1990)
3. Bron, C., Kerbosch, J.: Finding all cliques of an undirected graph. Commun. ACM **16**, 575–577 (1973)
4. Coie, J.D., Dodge, K.A., Coppotelli, H.: Dimensions and types of social status in the school. A cross-age comparison. Dev. Psychol. **18**, 557–570 (1982)
5. Coie, J.D., Dodge, K.A., Kupersmidt, J.B.: Peer group behaviours and social status. In: Asher, S.R., Coie, J.C. (eds.) Peer Rejection in Childhood, pp. 17–59. Cambridge University Press, New York (1990)
6. Dodge, K.A., Schlundt, D.C., Schocken, I., Delugach, J.D.: Social competence and children's sociometric status. The role of peer group entry strategies. Merrill Palmer Q. **29**, 309–336 (1983)
7. Giordano, G., Vitale, M.P.: On the use of external information in social network analysis. Adv. Data Anal. Classif. **5**(2), 95–112 (2011)
8. Goodreau, S., Kitts, J., Morris, M.: Birds of a feather, or friend of a friend? Using statistical network analysis to investigate adolescent social networks. Demography **46**, 103–125 (2009)
9. Hughes, C., Dunn, J.: Pretend you didn't know. Preschoolers? Talk about mental states in pretended play. Cogn. Dev. **12**, 381–403 (1997)
10. Lorrain, F., White, H.C.: Structural equivalence of individuals in social networks. J. Math. Sociol. **1**, 49–80 (1971)
11. Luce, R., Perry, A.: A method of matrix analysis of group structure. Psychometrika **14**, 95–116 (1949)
12. Maag, J.W., Vasa, S.F., Reid, R., Torrey, G.K.: Social and behavioural predictors of popular, rejected and average children. Educ. Psychol. Meas. **55**, 196–205 (1995)
13. Newcomb, A.F., Bukowski, W.M., Pattee, L.: Children's peer relations. A meta-analytic review of popular, rejected, neglected, controversial, and average sociometric status. Psychol. Bull. **113**, 99–128 (1993)
14. Perner, J., Ruffman, T., Leekam, S.R.: Theory of mind is contagious: you catch it from your sibs. Child Dev. **65**, 1228–1238 (1994)
15. Snijders, T.A.B., Pattison, P.E., Robins, G.L., Handcock, M.S.: New specifications for exponential random graph models. Sociol. Method. **36**, 99–153 (2006)
16. White, D.R., Reitz, K.P.: Graph and semigroup homomorphisms on networks of relations. Soc. Netw. **6**, 193–234 (1983)

# MCA-Based Community Detection

**Carlo Drago**

**Abstract** In this work, we propose a new approach for consensus community detection based on MCA. The advantage of this approach is synthetizing the information coming from different methods and secondarily to obtain for each node relevant evidence about their different classification on more communities. This result can be important because the position of the single node can be interpreted differently from the other nodes on the community. In this way, it is possible to identify also different roles of the communities inside the network. The approach is presented and is shown by considering simulated networks and, at the same, time by considering some real cases of networks. In particular, we consider the real network related to the Zachary Karate Club.

**Keywords** Social network analysis · Community detection · Consensus community detection · Communities · Multiple correspondence analysis

## 1 Community Detection Algorithms

This work is aimed at assessing the results obtained by different community detection algorithms on a complex network characterized by the hierarchical and the community structure [8]. The nodes on a community tend to be highly connected to each other and are connected sparsely with nodes of different communities. The aim of the community detection algorithms is to partition the network nodes into subsets. Different methodologies and algorithms tend to produce different results [13]. In this sense, it is necessary to consider an ensemble of methodologies in order to synthesize the results obtained [11]. At the same time, stochastic elements contained in the community detection algorithm can lead to different partitions when we consider different runs of the same algorithm [16]. In order to measure the level of agreement of the results, it is necessary to consider an ensemble approach in which, from the different results of the various algorithms, we can obtain a specific measure of robustness of the results. An approach in this sense is followed by [5, 6, 11] which

C. Drago (✉)
University of Rome 'Niccolo Cusano', Via Don Carlo Gnocchi 3, Rome, Italy
e-mail: carlo.drago@unicusano.it

 
F. Mola et al. (eds.), *Classification, (Big) Data Analysis and Statistical Learning*, Studies in Classification, Data Analysis, and Knowledge Organization,
https://doi.org/10.1007/978-3-319-55708-3_7

used an ensemble approach to combine the information from different methods. The novelty of this approach is the use of the MCA to obtain a synthesis of the results returned by the different algorithms.

## 2 MCA Based Consensus Community Detection

A network $G$ can be denoted as $G = (M, N)$ where $n$ is the number of nodes of the network and $M$ defines the presence of the number of the links connecting two nodes $k$ and $l$. In this sense, the network $G$ is based on the set of nodes and the set of edges which connect the different nodes. In particular, the edge $m_{k,l}$ defines the relationship between two nodes $n_k$ and $n_l$. Given an adjacency matrix $G = (G_{k,l})$, we can have 0 and 1: in the first case, there is no edge between $k, l$; in the second case, there exists a connection. It is possible to consider a grouping of the nodes on a partition $P$. The nodes are the different elements of the partitions $p_i$. The modularity $Q$ can be considered as a measure of quality of a partition of the network $G$. A higher modularity means a stronger intra-class structure with respect to a random network. The modularity optimization is particularly relevant for community detection problems. The aim of the optimization algorithms is to determine the best partitions in terms of modularity. In particular, this criterion is used to compare different partitions in order to obtain the optimal one. In this way, we can select the optimal number of partition for various community detection methods. But the modularity maximization can be problematic [10].

Hereinafter, we propose a different way to detect network communities considering an ensemble of different methods. We start considering different community detection methods and algorithms in order to capture different aspects of the original network $G$. In the second step, we define the communities taking into account and combining the results obtained from the different approaches. In the ensemble, we consider some known community detection methods [8, 17]: edge betweenness community, walktrap community, fastgreedy community, spinglass community, leading eigenvector community, infomap community, and label propagation. We obtain for each node a vector of membership where each element is the single partition or network community detected. Each partition represents a different network community. By considering the results in terms of memberships obtained for each node by each different methods (in columns), we can obtain the membership matrix $E$. From the different community detection methods considered, we obtain also the consensus matrix $A = [a_{i,j}]$ for each node $i, j = 1 \ldots n$. Each element in the cell of the matrix $A$ measures the consensus between the different methods on the two nodes. Each value of the cell in the matrix $A$ can have a value with a range from 0 to 1 (perfect consensus means that the couple of nodes $i, j$ are members of the same communities on all methods). When the consensus is perfect in the different methods, we obtain 1 for all values. At this point, we start from the membership matrix $E$, and then we extract and cluster the relevant factors. In particular, we use the multiple correspondence analysis (MCA).

**Algorithm 1** MCA-Based Community Detection

```
procedure
    while considering all community detection algorithms do
        perform the single community detection algorithm
        obtain a membership column for each method
    obtain the membership matrix E
    obtain the consensus matrix A
    perform the MCA from E
    perform the HCA
    cutting the dendrogram
    obtain the final membership vector
```

Each axis can be interpreted as the pattern for each method of community memberships. The information obtained in that way is also important because it allows to compare the different results obtained by the algorithms. Finally, by considering the majority of the numbers of communities extracted from each algorithm, we obtain the range of specific solutions to explore in the hierarchical clustering (HCA). In this way, we cut the dendrogram in order to detect the final communities. So we can analyze the results obtained using different numbers of communities (and the stable structures we are able to identify). Finally, we consider the different communities obtained by the procedure and we compare the membership using the Rand index (see [5]). The procedure can be depicted on the Algorithm 1.

## 3 The Analysis of the Consensus Matrix

We start the analysis from the consensus matrix. In this sense, we can observe that the matrix contains relevant information on the consensus of the different methods and algorithms on the different communities. In particular, it is possible to observe that each method can lead to different results. Each method can consider the different nodes in different communities [13]. The patterns can be analyzed by considering the multiple correspondence analysis. So it is relevant to analyze the different nodes which tend to be classified in the same communities by the same methods and at the same time the nodes which tend to be part of different communities. In this case, these nodes represent some statistical units which are critical because they are part of different communities w.r.t. different methodologies. In order to do that, we start from the $A$ matrix and we compute the different means from the different consensus from each algorithm or method of community detection. In this case, this measure can be computed by row or by column and can indicate the level of average consensus which it is possible to obtain for the considered node. In this sense, we consider the consensus related to the single node with all the different nodes of the network. It is possible to observe that the higher the consensus is, the higher probability of a node to be part of a structured group of nodes will be. These nodes are detected by different algorithms of community detection, each one considering a single community.

The nodes with lower values tend to be part of different communities and they tend to be members with other nodes of these communities. Thus, in this sense, the different methods or algorithms tend to have different results. From an interpretative point of view, this is an interesting result because these nodes tend to lead to new information.

## 4 Simulation Study

In order to experiment the different results obtained by the algorithm using different network structures, we consider some simulation experiments using synthetic data. In particular, we run the different algorithms on different network typologies and we perform the consensus approach in order to detect the different communities. The software used is R and in particular the package Igraph [4]. The different network typologies are for instance: the Barabasi model [1], the random graph model [7], and the forest fire model [12]. Then, we consider and compare the different results obtained by the different algorithms with the solution obtained from the consensus. The results shows that we obtain a reasonable synthesis from the different algorithms and in particular, we are able to see the different communities. In particular, we observe the results from the simulation on the Barabasi Model. The final results show six communities from the algorithm proposed. In particular, the final partition shows a Rand index of 0.9–1 with the different partitions obtained with the initial community detection methods. This result could be considered interesting because it is able to identify nodes which vary their membership from one community to another, due, for example, to a lower number of communities (or different type of communities). Usually, those nodes which have an higher centrality tend to be part of different communities. In this sense, we observe the different values for each node obtained by the consensus score. In particular, we can observe, for example, the value of the node 38 which shows a value of 0.10 (1 is the maximum) where the node 7 obtains a value of 0.22. In this case, the node 7 is more central and so tends to be involved in different communities with different nodes by considering the different methods. We consider at the same time, other experiments (see Table 1 for the results) focused on the understanding of the different results which can be achieved by considering different network structures. The experiments were organized as follows: we consider two families of network structures, the first one based on a simple structure which shows low density (based on a Barabasi model) and another one based on higher density networks than the previous (based on an Erdos Renyi model [7]). At the same time, we consider also families of networks based on a lower number of nodes a and a second one based on a higher one (40 vs. 100 nodes). The results are presented in Table 1. In particular, in Table 1, we observe the different results from the performed experiments in terms of Rand index. In this sense, the Rand index measures the level of similarity between the output obtained with the consensus and the five different methods: edge betweenness [9], walktrap [15], greedy optimization [3], the approach leading eigenvector [14], and finally the multilevel optimization of the modularity [2]. The number of communities chosen for the consensus is the mean of the different communities obtained by each method. We define, at the same time, the higher number of communities detected by all the different methods. By analyzing the results from Table 1, we can observe which increasing of the size of the network increases the number of communities detected at the same time. Simultaneously, by considering experiments with a higher density of the network, we can observe a decrease of the Rand index. In particular, if we consider an experiment in which a network is based on a higher number with a low density then that characteristic does not tend to change the results obtained using a lower number nodes in networks. In these cases, the network is well partitioned and the Rand index shows a reasonable value close to 1 showing an adequacy of the final solution with the different methods which are part of the methods used on the consensus. The results do not change greatly if we consider a higher level of density on the networks (there is a decrease of the Rand index). In these cases, the level of the agreement of the different community detection algorithms tends to decrease because the distinct algorithms and methodologies tend to find different communities but at the same time, we are able to identify some stable communities by considering the consensus of methodologies.

**Table 1** Experiments performed on higher network dimension and density level

| Mean | Max | Edge betweenness | Walktrap | Greedy optimization | Leading eigenvector | Multilevel | Density | Dimension |
|---|---|---|---|---|---|---|---|---|
| 5 | 8 | 0.85 | 0.90 | 0.89 | 0.83 | 0.86 | Higher | Lower |
| 6 | 8 | 0.83 | 0.87 | 0.88 | 0.79 | 0.92 | Higher | Lower |
| 6 | 9 | 0.87 | 0.77 | 0.84 | 0.85 | 0.79 | Higher | Lower |
| | | **0.85** | **0.85** | **0.87** | **0.82** | **0.86** | | |
| 7 | 8 | 0.99 | 0.99 | 0.99 | 0.98 | 0.99 | Lower | Lower |
| 7 | 9 | 0.98 | 0.96 | 0.98 | 0.98 | 0.98 | Lower | Lower |
| 7 | 9 | 0.99 | 0.97 | 0.99 | 0.98 | 0.99 | Lower | Lower |
| | | **0.99** | **0.97** | **0.99** | **0.98** | **0.99** | | |
| 11 | 48 | 0.88 | 0.88 | 0.84 | 0.78 | 0.87 | Higher | Higher |
| 9 | 37 | 0.81 | 0.86 | 0.80 | 0.80 | 0.82 | Higher | Higher |
| 9 | 39 | 0.87 | 0.81 | 0.81 | 0.76 | 0.85 | Higher | Higher |
| | | **0.86** | **0.85** | **0.82** | **0.78** | **0.85** | | |
| 13 | 20 | 0.96 | 0.92 | 0.97 | 0.96 | 0.97 | Lower | Higher |
| 14 | 21 | 0.94 | 0.93 | 0.94 | 0.95 | 0.94 | Lower | Higher |
| 12 | 18 | 0.95 | 0.93 | 0.95 | 0.92 | 0.95 | Lower | Higher |
| | | **0.95** | **0.92** | **0.96** | **0.95** | **0.96** | | |

*Notes* From the left, the columns are related to the number of the mean of the communities detected, the maximum, the different Rand index obtained by the partitions obtained with the consensus and the different methods considered: edge betweenness [9], walktrap [15], greedy optimization [3], leading eigenvector [14], and multilevel optimization of the modularity [2]. In bold is the mean of the Rand index obtained by each group of experiments. The last two columns are related to the network characteristics (density and dimension)

## 5 Application on Real Data

In order to test the approach with real data, we consider the known network related to the Zachary Karate Club (Fig. 1). The data was collected by Zachary in 1977 [18] from the members university karate club. In this sense, each edge represents a connection between two club members. At the same time, each specific node of the network represents a member of the Karate club. In order to apply the approach on the data, we consider the different algorithms and methods to the data, then we obtain the consensus matrix which indicates the different consensus scores which is related to each pair of nodes. In this sense, we can obtain different results by considering the different algorithms of community detection. Then, we apply on the results of the different community detection methodologies the multiple correspondence analysis. Finally, we can apply the cluster analysis from the results obtained and in this way

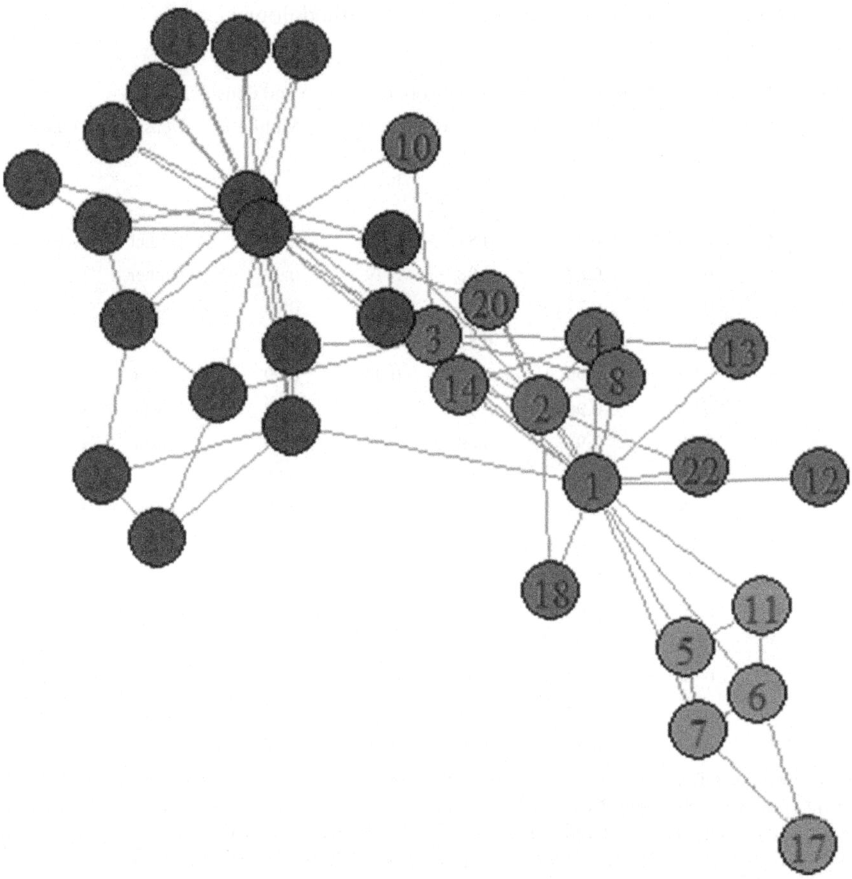


**Fig. 1** Results from the Kachary Club Network

deciding by majority the number of communities of the network (we consider the number of communities indicated by the majority of the algorithms of community detection), we can obtain the different memberships of the node to the different communities. Finally, we can compute the Rand index to the different results obtained with the initial algorithms in order to compare the results obtained from the partitions. In order to determine the nodes which are more interesting, as they belong to different communities, we compute the means for the columns from the consensus matrix. At the end of the analysis, we are able to identify three communities. The comparative analysis using the Rand index shows that the solution found is a synthesis of the different methods considered of community detection (the value of the Rand index in this sense is around 0.8–0.9 out of 1). The consensus score for each node shows that the nodes 5, 6, 7, 11, and 17 are on a particular position. It is possible to observe that they belong to different isolated communities on different methods than other nodes of the network.

## 6 Conclusions

In this work, we have considered a new algorithm which starting from a given network identifies the communities by considering a consensus approach. In this way, we can take in to account the information related to different methodologies and approaches and finally combine the information given from each method. In this sense, the proposed approach considers the same advantages of existing approaches which use consensus community detection. The main advantage here, differently from similar approaches, is the idea that the different nodes can be analyzed in order to specifically observe if a single node is a member for all the considered different algorithms of the same communities (and so it is a member of the same group of nodes). In this sense, the final idea is to discriminate the different nodes which can occur on networks in robust structures which repeat different methods and communities which appear for some configurations. The simulations and the application show the capacity to identify the critical nodes but at the same time the structure of the network which comes from the consensus community detection.

**Acknowledgements** I would like to thank Professor Carlo Lauro for the valuable discussion and comments. Any remaining errors are mine.

## References

1. Barabasi, A.L., Albert, R.: Emergence of scaling in random networks. Science **286**, 509–512 (1999)
2. Blondel, V.D., Guillaume, J.L., Lambiotte, R., Lefebvre, E.: Fast unfolding of communities in large networks. J. Stat. Mech. Theory Exp. **10**, P10008 (2008)
3. Clauset, A., Newman, M.E., Moore, C.: Finding community structure in very large networks. Phys. Rev. E **70**(6), 066111 (2004)

4. Csardi, G., Nepusz, T.: The Igraph software package for complex network research. Int. J. Complex Syst. **1695**. http://igraph.org (2006)
5. Drago, C., Balzanella, A.: Nonmetric MDS consensus community detection. In: Morlini, I., Minerva, T., Vichi, M. (eds.) Advances in Statistical Models for Data Analysis, pp. 97–105. Springer, Berlin (2015)
6. Drago, C., Cucco, I.: Robust communities detection in joint-patent application networks. https://works.bepress.com/carlo_drago/95/ (2013)
7. Erdos, P., Renyi, A.: On random graphs. Publicationes Mathematicae **6**, 290–297 (1959)
8. Fortunato, S.: Community detection in graphs. Phys. Rep. **486**(3), 75–174 (2010)
9. Freeman, L.C.: Centrality in social networks I: conceptual clarification. Soc. Netw. **1**, 215–239 (1979)
10. Good, B.H., de Montjoye, Y.A., Clauset, A.: Performance of modularity maximization in practical contexts. Phys. Rev. E **81**(4), 046106 (2010)
11. Lancichinetti, A., Fortunato, S.: Community detection algorithms: a comparative analysis. Phys. Rev. E **80**(5), 056117 (2009)
12. Leskovec, J., Kleinberg, J., Faloutsos, C.: Graph evolution: densification and shrinking diameters. ACM Trans. Knowl. Discov. Data (TKDD) **1**(1), 2 (2007). https://doi.org/10.1145/1217299.1217301
13. Leskovec, J., Lang, K.J., Mahoney, M.: Empirical comparison of algorithms for network community detection. In: Proceedings of the 19th international Conference on World Wide Web, ACM, pp. 632–640 (2010)
14. Newman, M.E.J.: Finding community structure using the eigenvectors of matrices. Phys. Rev. E **74**, 036104 (2006)
15. Pons, P., Latapy, M.: Computing communities in large networks using random walks. J. Graph Algorithms Appl. **10**(2), 191–218 (2006)
16. Treviño, S., Sun, Y., Cooper, T.F., Bassler, K.E.: Robust detection of hierarchical communities from $Escherichiacoli$. Gene expression data. PLoS Comput. Biol. **8**, e1002391 (2012)
17. Yang, Z., Algesheimer, R., Tessone, C.J.: A Comparative analysis of community detection algorithms on artificial networks. Sci. Rep. **6**, 30750, EP (2016)
18. Zachary, W.W.: An information flow model for conflict and fission in small groups. J. Anthropol. Res. **33**(4), 452–473 (1977)

# Part III
# Exploratory Data Analysis

# Rank Properties for Centred Three-Way Arrays

**Casper J. Albers, John C. Gower and Henk A. L. Kiers**

**Abstract** When analysing three-way arrays, it is a common practice to centre the arrays. Depending on the context, centring is performed over one, two or three modes. In this paper, we outline how centring affects the rank of the array; both in terms of maximum rank and typical rank.

**Keywords** Three-way analysis · Multiway analysis · Maximum rank · Typical rank · CANDECOMP/PARAFAC

## 1 Introduction

Let $\underline{\mathbf{X}}$, of dimension $I \times J \times K$, be a three-way array (also termed a tensor) with entries $x_{ijk}$. For the sake of simplicity, we assume that $I \leq J \leq K$ (whenever this is not the case, we can make this the case without loss of generality by simply permuting the labels of the array).

In the analysis of arrays, the concept of rank is of importance, for the same reasons why it is important in the analysis of a two-way data matrix. The rank of a matrix is the dimension of the vector space spanned by its columns, i.e. the maximum number of distinct components the array can be decomposed into. For arrays, the concept of rank is similar, but now for three dimensions rather than two. (See Sect. 2 for details.)

C. J. Albers (✉) · H. A. L. Kiers
Department of Psychometrics & Statistics, University of Groningen, Groningen, The Netherlands
e-mail: c.j.albers@rug.nl

H. A. L. Kiers
e-mail: h.a.l.kiers@rug.nl

J. C. Gower
Department of Mathematics & Statistics, The Open University, Milton Keynes, UK
e-mail: john.gower@open.ac.uk

 
F. Mola et al. (eds.), *Classification, (Big) Data Analysis and Statistical Learning*, Studies in Classification, Data Analysis, and Knowledge Organization,
https://doi.org/10.1007/978-3-319-55708-3_8

In this paper, we study the consequences of centring, over either one, two or three modes, on the rank of the array. Centring three-way arrays is common practice in data analysis; similar to the centring of data matrices prior to performing a principal components analysis.

One should distinguish different types of pre-scaling data. One purpose of pre-scaling is (i) to reduce the effects of incommensurabilities in different parts of the data, or transformations to more acceptable measures such as logs or square roots, but another is (ii) to isolate different substantive components which deserve separate examination. Normalisation in principal component analysis is an example of (i), while removing the mean is an example of (ii). In this paper, we are concerned with (ii) and note that the separate components of analysis not only enhance interpretation but may also reduce rank. Thus, although centring is usually performed solely to improve model fit, e.g. of a CANDECOMP/PARAFAC or Tucker3 decomposition, it is important to realise that centring can have a substantive effect. In the analysis of additive models, especially when studying interactions [1, 2], it is common to partition $\underline{\mathbf{X}}$ into parts for the overall mean, main effects, biadditive effects and triadditive effects:

$$x_{ijk} = m + \{a_i + b_j + c_k\} + \{d_{jk} + e_{ik} + f_{ij}\} + g_{ijk}, \tag{1}$$

where the terms with a single suffix represent main effects, those with double suffices two-factor interactions and $g_{ijk}$ represent contributions from three-factor interactions. Some components of the interactions may be regarded as 'error'. The defining equations are subsumed in the identity:

$$\begin{aligned}\hat{x}_{ijk} = {} & x_{...} + \{(x_{i..} - x_{...}) + (x_{.j.} - x_{...}) + (x_{..k} - x_{...})\} \\ & + \{(x_{.jk} - x_{.j.} - x_{..k} + x_{...}) + (x_{i.k} - x_{i..} - x_{..k} + x_{...}) \\ & + (x_{ij.} - x_{i..} - x_{.j.} + x_{...})\} \\ & + (x_{ijk} - x_{.jk} + x_{i.k} + x_{ij.} + x_{i..} + x_{.j.} + x_{..k} - x_{...}),\end{aligned} \tag{2}$$

where the expressions in parentheses in (2) estimate the corresponding parameters in (1).

The triadditive model for given choices of $P \le I$, $Q \le J$, $R \le K$ and $S$ is given by

$$x_{ijk} = m + a_i + b_j + c_k + \sum_{p=1}^{P} d_{jp}\tilde{d}_{kp} + \sum_{q=1}^{Q} e_{iq}\tilde{e}_{kq} + \sum_{r=1}^{R} f_{ir}\tilde{f}_{jr} + \sum_{s=1}^{S} g_{is}\tilde{g}_{js}\tilde{\tilde{g}}_{ks} + \varepsilon_{ijk} \tag{3}$$

(By taking $S = 0$, one obtains the biadditive model.) To make this model identifiable, zero-sum identification constraints are required. Without such constraints, exactly the same fit would be obtained if, e.g. a nonzero value $\varepsilon$ was added to all $a_i$ and subtracted from all $b_j$. Requiring zero-sums is in line with the concept of marginality [3], i.e. first fitting an overall effect, then main effects on the residuals, then biadditive effects on the residuals and so on. In biadditive models, zero-sum constraints are straightforward, but this is not the case in triadditive models since for triadditive models, some forms of centring change the form of the model. One consequence is that the least-squares estimates of the triadditive interaction parameters depend on how exactly, i.e. by how many components, each of the biadditive terms is modelled [2, 4]. To bypass these issues, one may fit the triadditive part conditional on the main effects and the saturated biadditive components of the model. That is, we fit the triadditive part of the model to the biadditive residual table:

$$z_{ijk} = x_{ijk} - x_{\cdot jk} - x_{i\cdot k} - x_{ij\cdot} + x_{i\cdot\cdot} + x_{\cdot j\cdot} + x_{\cdot\cdot k} - x_{\cdot\cdot\cdot}. \tag{4}$$

Triadditive interactions in (3) may be modelled using a truly triadic model such as the CANDECOMP algorithm [5], minimising

$$\sum_{i,j,k,r} (z_{ijk} - a_{ir}b_{jr}c_{kr})^2 \tag{5}$$

(see next section).

Thus, centring over one or two modes can be seen as taking out main effects or two-way interactions, respectively, and analyse them separately. It is important to wonder whether it is sensible for the problem at hand to perform the chosen type of centring. In the words of [6]: 'It is important that the final model or models should make sense physically: at a minimum, this usually means that interactions should not be included without main effects nor higher degree polynomial terms without their lower-degree relatives.'

In this paper, we study the effect of various types of centring on the rank of three-way arrays. This paper is organised as follows. In Sect. 2, we establish notation and recall relevant definitions from literature. Section 3 hosts the main theorem on the rank properties of centred arrays. We conclude with a series of examples in Sect. 4.

## 2 Notation and Known Results

We adhere to the standardised notation and terminology as proposed by [7]. The mode A matricised version of $\underline{\mathbf{X}}$ is given by the $I \times JK$ matrix $\mathbf{X}_a$ with all vertical fibres of a three-way array collected next to each other. Mode B and mode C matricised versions are defined in analogous ways. The vectorisation operator vec implies columnwise vectorisation and $\otimes$ is used for the Kronecker product. Furthermore, array $\underline{\mathbf{G}}$ is the so-called superidentity core array with elements $g_{pqr} = 1$ if $p = q = r$ and $g_{pqr} = 0$ otherwise. Finally, $\mathbf{I}$ is the identity matrix and $\mathbf{0}$ and $\mathbf{1}$ are column vectors with all values either 0 or 1, respectively, all of accommodating size.

There is a considerable literature on the ranks of general three-way arrays, summarised by [8], [9, Sect. 2.6] and [10, Sect. 8.4]. There are two types of rank to be considered: maximum rank and typical rank.

**Definition 1** The *maximum rank* of three-way array $\mathbf{X}$, with dimension $I \times J \times K$, is defined as the smallest value of $R$ that can give exact fit for

$$\sum_{i,j,k=1}^{I,J,K} \sum_{r=1}^{R} \left(z_{ijk} - a_{ir}b_{jr}c_{kr}\right)^2 . \tag{6}$$

**Definition 2** The *typical rank* is defined by [8, p. 3] as follows: 'The typical rank of a three-way array is the smallest number of rank-one arrays that have the array as their sum, when the array is generated by random sampling from a continuous distribution.'

An earlier definition of typical rank by [11] is given in a more complicated way [8], but on [11, p. 96] (bottom paragraph) seems to converge to Ten Berge's definition. So we follow the latter one. Since typical rank can be smaller than maximal rank (see [8] for examples), it will be of more practical usefulness than maximal rank, as this already provides a practical upper bound to the number of components one wants to decompose the array in.

When $J$ is small (close to $I$), the rank of $\underline{\mathbf{X}}$ is less than the upper bound $K$ but it seems to coincide with the upper bound when $K \geq IJ$. These results are less simple than those for matrices, but have in common more concern with good low-rank approximations to (6) rather than with the rank itself. The three-way interaction in (4) is free both of main effects and of two-way interactions, and so all its margins are null. Thus, the three-way table $\mathbf{Z} = \{z_{ijk}\}$ is a special form of a triadditive table and it may be expected to have special properties. In particular, we may expect it to have lower triadditive rank than for unconstrained triadditivity. Also, when only some of the modes are centred, the rank is expected to be reduced. A formal result that establishes this expectation is given in the following section.

## 3 Main Result

**Theorem 1** *Let the class of real-valued three-way arrays $I \times J \times K$ have at most maximum rank $f(I, J, K)$, where $f(I, J, K)$ denotes a particular function of $I$, $J$ and $K$. Then, a three-way array obtained by centring an array from this class of arrays will have rank at most equal to $f(I^*, J^*, K^*)$, where the starred versions denote $(I-1)$ or $I$, $(J-1)$ or $J$, $(K-1)$ or $K$, respectively, depending on whether or not the array has been centred across the first, second and/or third mode, respectively.*

Before we prove Theorem 1, we make three remarks.

*Remark 1* It should be mentioned that in [12, p. 375] it was already stated that double centring of symmetric matrices 'has a rank-reducing impact on the symmetric array' and they give a concise proof for that. The above Theorem follows the same reasoning as [12] but gives a more general result.

*Remark 2* We conjecture that the analogous theorem where 'maximal rank' is replaced by 'typical rank' also holds. For several classes of arrays of size $I \times J \times K$, the typical rank has been given as a function $f(I, J, K)$ of $I$, $J$ and $K$, and our conjecture is that like for the maximal rank, upon centring the array across the first, second and/or third mode, the typical rank should be given by $f(I^*, J^*, K^*)$, where the starred versions denote $(I-1)$ or $I$, $(J-1)$ or $J$, and $(K-1)$ or $K$, respectively, depending on whether or not the array has been centred across the first, second and/or third mode, respectively. In fact, [12] apply this reasoning. This may very well be correct, but we do not know whether we can still consider a class of random arrays which (all in the same way) have been double centred and from which two slices have been chopped off as 'generated by random sampling from a continuous distribution'.[1]

*Remark 3* We have no knowledge of any encompassing function $f(I, J, K)$ describing the maximal rank of $I \times J \times K$ arrays, but there are results for some general classes of $I \times J \times K$ arrays for the maximal or typical rank (see below), for example, $f(I, J, K) = I$ for all arrays for which $JK - J < I < JK$, and f now denotes typical rank [13]. However, in many cases, no results are less general, and the function $f$ in fact refers to a partially known mapping of the set $\{I, J, K\}$ on the real field $\mathbb{R}$. The mapping can be deduced from the literature, the latest summary of which (to our knowledge) has been given by [8].

*Proof* (**of Theorem 1**) Recall that the maximum rank of a three-way array $\underline{\mathbf{X}}$ is given by the smallest number $R$ for which for all $i, j, k$ it holds that $x_{ijk} = \sum_{r=1}^{R} a_{ir} b_{jr} c_{kr}$. In matrix notation, this is

$$\mathbf{X}_a = \mathbf{A}\mathbf{G}_a(\mathbf{C} \otimes \mathbf{B})', \tag{7}$$

where $\mathbf{X}_a$ and $\mathbf{G}_a$ denote the A-mode matricised versions of $\mathbf{X}$ and $\mathbf{G}$, respectively and $\mathbf{A}(I \times R)$, $\mathbf{B}(J \times R)$ and $\mathbf{C}(K \times R)$ denote the component matrices for the three modes. The following equivalent expressions can be given upon B- or C-mode matricisation:

$$\mathbf{X}_b = \mathbf{B}\mathbf{G}_b(\mathbf{A} \otimes \mathbf{C})', \tag{8}$$

[1]Technically, this is a matter of assessing the class' Lebesgue measure, to which we have no clue. To give an example that generally performed transformations may alter 'randomness' properties, consider for instance squaring all values, which clearly affects the Lebesgue measure of subclasses of the class of such arrays. However, because [12]'s transformations, as our own, are rank preserving, we expect that the results that are only proven for the maximal rank, also hold for the typical rank of classes of arrays.

and

$$\mathbf{X}_c = \mathbf{C}\mathbf{G}_c(\mathbf{B} \otimes \mathbf{A})'. \tag{9}$$

Obviously,

$$\mathbf{X}_a = \mathbf{A}\mathbf{G}_a(\mathbf{C} \otimes \mathbf{B})' \quad \text{iff} \quad \mathbf{S}\mathbf{X}_a(\mathbf{U} \otimes \mathbf{T})' = \mathbf{S}\mathbf{A}\mathbf{G}_a(\mathbf{U}\mathbf{C} \otimes \mathbf{T}\mathbf{B})', \tag{10}$$

for any nonsingular square matrices **S**, **T** and **U**. Now suppose that $\underline{\mathbf{X}}$ is centred across mode A, then for the vector $\mathbf{u} = (1, 1, \ldots, 1)'$ it holds that

$$\mathbf{u}'\mathbf{A}\mathbf{G}_a(\mathbf{C} \otimes \mathbf{B})' = \mathbf{0}'. \tag{11}$$

Choosing **S** as a nonsingular matrix the first $I - 1$ rows of which are not centred (e.g. by taking these equal to the first $I - 1$ rows of the $I \times I$ identity matrix) and the last row is the vector $\mathbf{u}'$. Then, the last row of **SA** and hence of

$$\mathbf{S}\mathbf{X}_a = \mathbf{S}\mathbf{A}\mathbf{G}_a(\mathbf{C} \otimes \mathbf{B})' \tag{12}$$

equals $\mathbf{0}'$. Thus, the matricised array $\mathbf{S}\mathbf{X}_a$ can be written as $\begin{pmatrix} \mathbf{Y}_a \\ \mathbf{0} \end{pmatrix}$, in other words, as the concatenation of the $(I - 1) \times J \times K$ array $\mathbf{Y}_a$ containing the first $I - 1$ rows of $\mathbf{S}\mathbf{X}_a$ and the vector $\mathbf{0}$. For array $\underline{\mathbf{Y}}$, written in matricised form $\mathbf{Y}_a$, it holds that it has rank at most equal to $f(I - 1, J, K)$. Hence, it has a decomposition as in (7) for $R = f(I - 1, J, K)$. As a consequence, $\mathbf{S}\mathbf{X}_a$ can be written as

$$\begin{pmatrix} \mathbf{Y}_a \\ \mathbf{0} \end{pmatrix} = \begin{pmatrix} \mathbf{A}^*\mathbf{G}_a(\mathbf{C} \otimes \mathbf{B})' \\ \mathbf{0} \end{pmatrix} = \begin{pmatrix} \mathbf{A}^* \\ \mathbf{0} \end{pmatrix} \mathbf{G}_a(\mathbf{C} \otimes \mathbf{B})',$$

where $\mathbf{A}^* = \mathbf{S}\mathbf{A}$ and thus $\mathbf{S}\mathbf{X}_a$ has a decomposition in $R = f(I - 1, J, K)$ components. As a consequence, because of (10), also $\mathbf{X}_a$ has a decomposition in $R = f(I - 1, J, K)$ components, from which it follows immediately that $\mathbf{X}_a$ has at most rank $f(I - 1, J, K)$.

This concludes the proof of the theorem for centring across mode A. Centring across mode B or C can be proven completely analogously, using matricised forms (8) and (9). □

## 4 Examples

In this section, we give a few examples.

*Example 1* $100 \times 3 \times 2$ arrays.
The theorem could be seen as stating that centring across one mode will always reduce the maximal rank of a class of arrays by a factor $(G - 1)/G$ where G denotes

$I$, $J$ or $K$ depending on the mode across which we centre. This, however, need not be true, as is obvious in the case where $I \gg JK$. Suppose we deal with the class of $100 \times 3 \times 2$ arrays. Then, the typical rank will at most be 6 [8]. In this case, the rank does not depend on $I$ at all (since $I > JK$). Hence, centring across mode A, will lead to $R = f(I-1, J, K)$, which also equals 6 [8]. However, centring across mode B and C, does have an effect on the maximal rank. Provided that this is $JK = 6$, centring only across mode B reduces it to $(J-1)K = 2 \times 2 = 4$, centring only across mode C reduces it to $J(K-1) = 3 \times 1 = 3$ and centring across both modes reduces it to $(J-1)(K-1) = 2 \times 1 = 2$, a threefold reduction compared to the original typical rank.

*Example 2* $10 \times 4 \times 3$ arrays.
Following [8], for the class of arrays of size $10 \times 4 \times 3$, the typical rank is 10. Table 1 gives the typical rank for all combinations of centring of such arrays. Clearly, in this case, the effect of single centring depends on the mode that is centred (see rows 2–4 in the table). This is even more so for the effect of double centring (rows 5–7).

*Example 3* $2 \times J \times K$ arrays.
A third special case is concerned with triadditive interactions arrays, such as $\underline{\mathbf{Z}}$ as given in Eq. (4), with $I = 2$ and $J, K > 2$. In this case, the rank is $J - 1$ and there are various ways decomposing the array into three component matrices with perfect fit. A convenient decomposition is the following. As $\underline{\mathbf{Z}}$ has zero-sum marginals, it is clear that $\mathbf{A} \propto (\mathbf{1}, -\mathbf{1})'$ (with dimension $2 \times (J-1)$) and it is convenient to choose $\mathbf{A} \propto (\mathbf{1}, -\mathbf{1})'$. Then, the matrices $\mathbf{B}$ ($J \times (J-1)$) and $\mathbf{C}$ ($K \times (J-1)$) can be obtained from the $J \times K$ matrix $\mathbf{Z}_1 = -\mathbf{Z}_2$ through a singular value decomposition, where $\mathbf{Z}_1$ and $\mathbf{Z}_2$ denote the first and second horizontal slices of $\underline{\mathbf{Z}}$.

However, a simpler decomposition emerges upon writing

$$\mathbf{Z}_1 = \begin{pmatrix} \mathbf{Z}_1^* \\ -\mathbf{1}'\mathbf{Z}_1^* \end{pmatrix},$$

**Table 1** Example of effects of (combinations of) centring of modes of $10 \times 4 \times 3$ arrays

| Mode A | Mode B | Mode C | $I^* \times J^* \times K^*$ | Typical rank |
|---|---|---|---|---|
| N | N | N | $10 \times 4 \times 3$ | 10 |
| C | N | N | $9 \times 4 \times 3$ | 9 |
| N | C | N | $10 \times 3 \times 3$ | 9 |
| N | N | C | $10 \times 4 \times 2$ | 8 |
| C | C | N | $9 \times 3 \times 3$ | 9 |
| C | N | C | $9 \times 4 \times 2$ | 8 |
| N | C | C | $10 \times 3 \times 2$ | 6 |
| N | N | N | $9 \times 3 \times 2$ | 6 |

In the table, C means centring across that mode, and N means not centring across that mode. Results are derived from Table 1 from [8]. The lines separate no centring, single centring, double centring and triple centring where $\mathbf{Z}_1^*$ contains the first $J-1$ rows of $\mathbf{Z}$. Then, obviously, $\mathbf{Z}_1 = \mathbf{BC}'$, where $\mathbf{B} = (\mathbf{I}, -\mathbf{1})'$, with $\mathbf{I}$ of order $(J-1) \times (J-1)$, and $\mathbf{C}' = \mathbf{Z}_1^*$. As, clearly, $\mathbf{A}, \mathbf{B}$ and $\mathbf{C}$ all have $J-1$ columns, thus constituting a rank $J-1$ decomposition of $\underline{\mathbf{Z}}$. The convenience of this solution lies in that of the three component matrices, only $\mathbf{C}$ contains values that relate to the data itself.

## 5 Conclusion

To conclude, it has been seen that centring often, but not always reduced the rank of arrays. Sometimes, the reduction is dramatic, and comes close to practical values. For instance, a researcher should not be surprised to find perfect PARAFAC fit already for $R = 2$ when analysing a $100 \times 3 \times 2$ array which has been centred across B- and C-mode.

## References

1. Albers, C.J., Gower, J.C.: A contribution to the visualisation of three-way arrays. J. Multivar. Anal. **132**, 1–8 (2014)
2. Albers, C.J., Gower, J.C.: Visualising interactions in bi- and triadditive models for three-way tables. Chemometr. Intell. Lab. Syst. **167**, 238–247 (2017)
3. Nelder, J.A.: A Reformulation of linear models. J. Roy. Stat. Soc. Ser. A (General) **140**(1), 48–77(1977)
4. Gower, J.C..: The analysis of three-way grids. In: Slater, P. (ed.) Dimensions of Intra Personal Space. The Measurement of Intra Personal Space by Grid Technique, vol. 2, pp. 163–173. Wiley, Chichester (1977)
5. Carroll, J.D., Chang, J.J.: Analysis of individual differences in multidimensional scaling via an $n$-way generalization of 'Eckart-Young' decomposition. Psychometrika **35**, 283–319 (1970)
6. McCullagh, P., Nelder, J.A.: Generalized Linear Models, 2nd edn. Chapman & Hall/CRC, Boca Raton, Florida (1989)
7. Kiers, H.A.L.: Towards a standardized notation and terminology in multiway analysis. J. Chemometr. **14**, 105–122 (2000)
8. TenBerge, J.M.F.: Simplicity and typical rank results for three-way arrays. Psychometrika **76**, 3–12 (2011)
9. Smilde, A.K., Bro, R., Geladi, P.: Multi-way analysis with applications in the chemical sciences. Wiley, Hoboken, New Jersey (2004)
10. Kroonenberg, P.M.: Applied Multiway Data Analysis. Wiley, Hoboken, New Jersey (2008)
11. Lickteig, T.: Typical tensorial rank. Linear Algebra Appl. **69**, 95–120 (1985)
12. ten Berge, J.M.F., Sidiropoulos, N.D., Rocci, R.: Typical rank and Indscal dimensionality for symmetric threeway arrays of order $I \times 2 \times 2$ or $I \times 3 \times 3$. Linear Algebra Appl. **388**, 363–377 (2004)
13. ten Berge, J.M.F.: The typical rank of tall three-way arrays. Psychometrika **65**, 525–532 (2000)

# Principal Component Analysis of Complex Data and Application to Climatology

**Sergio Camiz and Silvia Creta**

**Abstract** For the study of *El Niño* phenomenon, winds data collected from the Equator belt of Pacific ocean would be analyzed through *PCA*. In this paper, the 2-dimensional nature of winds is discussed in respect to the possible ways in which *PCA* may be implemented. Among others, *complex PCA* is proposed and compared on a small example to other methods based on real *PCA*. Then, the first results on a larger data table are illustrated.

**Keywords** Principal component analysis · Complex principal component analysis · El Niño

## 1 Introduction

In this paper, we propose *Complex Principal Component Analysis* (*CPCA*, [1]) as a viable tool to analyze wind data time series in an exploratory framework to study *El Niño* phenomenon. *El Niño* is a large-scale oceanic warming in the tropical Pacific Ocean that occurs every few years. The Southern Oscillation is characterized by an interannual seesaw in tropical sea level pressure (*SLP*) between the western and eastern Pacific, consisting of a weakening and strengthening of the easterly trade winds over the tropical Pacific. A close connection was recognized between El Niño and the Southern Oscillation (*ENSO*) and they are two different aspects of the same phenomenon. *ENSO* is supposed to be caused by a positive ocean-atmosphere feedback involving the surface trade winds blowing from the east to the west across the tropical Pacific Ocean, the rising air in the tropical western Pacific, the upper-level winds blowing from the west to the east, and the sinking air returned back to the

S. Camiz (✉)
Dipartimento di Matematica, Sapienza Università di Roma, Rome, Italy
e-mail: sergio@camiz.net

S. Creta
Sapienza Università di Roma, Rome, Italy
e-mail: silviacreta@gmail.com



F. Mola et al. (eds.), *Classification, (Big) Data Analysis and Statistical Learning*, Studies in Classification, Data Analysis, and Knowledge Organization,
https://doi.org/10.1007/978-3-319-55708-3_9

surface in the tropical eastern Pacific. An initial positive sea surface temperature (*SST*) anomaly in the equatorial eastern Pacific reduces the east-west *SST* gradient and hence results in weaker trade winds around the equator. The weaker trade winds in turn drive the ocean circulation changes that further reinforce SST anomaly. This positive ocean-atmosphere feedback leads the equatorial Pacific to a warm state, i.e., the warm phase of *ENSO—El Niño*. A turnabout from a warm phase to a cold phase, has been named *La Niña* [2, 3].

Indeed, the immense amount of energy involved causes effects that go largely beyond the Pacific shores and affect most of the planed climate, at least at low/mid-latitudes. This led [4, 5] to start studying the phenomenon through exploratory Principal Component Analysis (*PCA*, [6]) based on climatic time series.

Time series is usually studied one by one via autoregressive models for forecasting purposes. In the case of climatological data recorded by a large number of stations, such a study could not lead to a global understanding of the phenomenon. This is the reason why the multidimensional approach is largely found in literature [7]. In this framework, [4] tried to understand first how the Pacific Ocean could be partitioned in homogeneous zones and achieved it by using a hierarchical factor classification, that produces at each step a representative time series. Once decided a partition, the representative time series of each zone may be modeled through autoregressive models. Indeed, methods may be found combining the two approaches: for the time dependence, different proposals may be found in [8–10].

Camiz et al. [4] started considering temperatures, aiming at broadening the study to both pressure and winds. In [5] the analysis of winds raised a problem, since, unlike the other scalar measures, they are vectors with two orthogonal components: zonal *West–East* and meridional *South–North* (*EW* and *SN*, respectively). In order to give sense to *PCA*, a relevant requirement is that the principal components should be themselves wind time series, thus 2-dimensional vectors with analogous decomposition.

Considering the two components data tables, several options may be possible to get *"principal" wind components*:

1. Perform a *PCA* of each table separately: this would give two unrelated representations, with different meaning, so that no "principal" winds would really result, unless pairing the corresponding principal components of the two analyses in joint plane representations.
2. Perform a *PCA* of the tables joint side by side: this would give joint principal components to which both tables would concur. Indeed, the principal components may be decomposed in two, each a linear combination of either tables columns, that would result in different weighting of the two subcomponents. Then, a graphical plane representation would result by considering the pair of subcomponents as orthogonal axes.
3. Perform a *PCA* of the stacked tables: this time two points would correspond to the same observation in the factor space. Indeed, splitting the principal components into two stacked subcomponents and using them as orthogonal axes allows a

representation of both rows and columns on the spanned plane. Nevertheless, the statistics would be pooled, with doubtful significance.

4. Consider the two tables as the real and imaginary part, respectively, of a complex data table and perform a complex *PCA* (*CPCA*) of this table. The first complex principal component is a plane, whose coordinates are the component's real and imaginary parts, respectively. This is possible since the correlation matrix is hermitian, with real eigenvalues, so that both complex singular value decomposition [11, 12] and Eckart and Young's theorem [13], thus *PCA*, may be applied. Indeed, the real and imaginary parts of complex principal components correspond to the "principal" winds zonal and meridional components, respectively.

In [5] the first three methods were investigated, based on the classic graphics on the first two axes, and the quoted drawbacks resulted evidently. Thus, the interpretation of the results appeared really difficult. In this paper we compare the results of the four methods applied to a small data set and then we apply *CPCA* to the wind data of the Equator belt of Pacific from 1991 to 2012. The results concern only the first *CPCA* dimension, that is a plane, in which the two real and imaginary parts of the coordinates correspond to the first principal component's zonal and meridional components. To get methods really comparable, instead of using the other methods the first two factors, as usual, we used the two first subcomponents, as described above.

## 2 Complex Singular Value Decomposition

Singular Value Decomposition (*SVD*, [12]) and Eckart and Young's theorem [13] may be applied to a complex matrix in analogous way as to a real one, so that *PCA*, essentially based on both, may be applied to a complex table too. The idea was first introduced by [11]; in the complex case, the theorems may be formulated as follows:

**Theorem 1** (Singular Value Decomposition) *Let a complex matrix* $\boldsymbol{A} \in \mathbb{C}^{m\times n}$, *then two unit matrices exist* $\boldsymbol{U} \in \mathbb{C}^{m\times m}$ *and* $\boldsymbol{V} \in \mathbb{C}^{n\times n}$ *such that*

$$\boldsymbol{U}^H \boldsymbol{A} \boldsymbol{V} = \boldsymbol{\Sigma} = \begin{pmatrix} diag(\sigma_1, \ldots, \sigma_p) & 0 \\ 0 & 0 \end{pmatrix}$$

*with* $\boldsymbol{\Sigma} \in \mathbb{R}^{m\times n}$, $p = rank(\boldsymbol{A}) \leq min(m, n)$ *and* $\sigma_1 \geq \sigma_2 \geq \ldots \geq \sigma_p > 0$.

**Theorem 2** (Eckart and Young) *Let* $\boldsymbol{A}$ *an* $n \times p$ *complex matrix with rank K and let* $\boldsymbol{A} = \boldsymbol{U}\boldsymbol{\Lambda}\boldsymbol{V}^H$ *its singular value decomposition, with* $\lambda_k$ *sorted in decreasing order. Then the rank* $k < K$ *matrix* $\boldsymbol{B}$ *such that* $\|\boldsymbol{A} - \boldsymbol{B}\|^2 = \min_{\mathbb{C}}(\|\boldsymbol{A} - \boldsymbol{C}\|^2)$ *is given by:*

$$\boldsymbol{B} = \boldsymbol{U}_{n,k}\boldsymbol{\Lambda}_k\boldsymbol{V}^H_{k,p}$$

*where $\boldsymbol{U}_{n,k}$ is formed by the first k columns of $\boldsymbol{U}$, $\boldsymbol{V}^H_{k,p}$ is formed by the first k rows of $\boldsymbol{V}^H$, and $\boldsymbol{\Lambda}_k$ is a diagonal matrix formed by the first k diagonal elements of $\boldsymbol{\Lambda}$.*

Due to loss of space, we refer to the demonstrations in [12].

## 3 Data

For the *NOAA TAO* project [14], the *NOAA* administration publishes on its web site (www.pmel.noaa.gov/tao) data daily measured on an array of 8 × 11 buoys regularly placed in the Equatorial belt of Pacific Ocean. For our work, we downloaded the two wind components at sea level, namely West–East (zonal component) and South–North (meridional component). Albeit the data are available starting March 1st, 1980, we limited the study period to the 22 years period 1991–2012, in which 68 buoys have been active, whereas only 27 provided data during the previous years. As well, we could not take into account the more recent data, since *NOOA* itself warns about its use, due to the loss of maintenance of the buoys system. Thus, we built a data tables with 68 columns (each buoy time series of observations) and 8036 rows (the daily observations along 22 years).

A special consideration deserves the treatment of missing data, since they are many in each buoy. In order to complete the series, we decided to estimate the missing values, but, instead of substituting them with the average of each time series, we decided to apply a methodology based on the Kohonen algorithm [15]. It consists in classifying statistical units, in our case the daily observations, according to either rectangular or hexagonal cells, also providing class centroids computed by the algorithm itself. As well, proximities among classes may be computed. In the case of missing data, the Kohonen algorithm assigns the same the corresponding observations to classes. Thus, the missing values may be estimated through the mean values (centroids) of the class of belonging.

To compare the four methods, we took into account only the 12 westernmost buoys of the array, during the 11-years period 2001–2011. The daily observations form two 4017 × 11 tables. All computations were done through *MATLAB* [16].

## 4 Results

### *4.1 The Comparison of Methods*

In Fig. 1(left) the 12 buoys' time series are represented by using as coordinates their correlations with the first principal wind components (zonal and meridional) issued by the four different analyses according to the given description. Note the different pattern in both separate (a) and side by side tables analyses (b) in respect to the others:

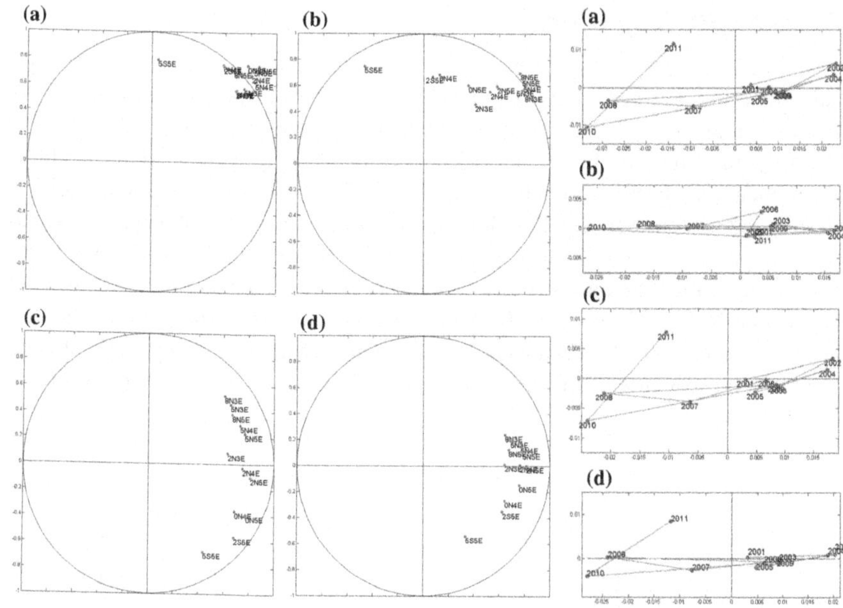


**Fig. 1** Left: the representation of the buoys' correlations with the first zonal and meridional wind principal components issued by the four *PCA* based methods. Right: the representation of the years on the planes spanned by the zonal and meridional wind components issued by the first principal component of each of the four *PCA*-based methods. **a** separate analyses; **b** tables side by side; **c** stacked tables; and **d** complex analysis

indeed, they measure correlations with the two components built independently, thus the corresponding points may be out of the circle of correlation. In both other analyses, points are distributed along the meridional component, with small variation along the zonal one, with the stacked analysis (c) more distributed than the complex (d) one. The latter circle of correlation may be read as a different composition of the inertia, higher for the zonal component for all but the $5S5E$ series, which is more developed along the meridional component. In Fig. 1(right), the pattern of the years, projected as supplemental characters on the principal wind plane, is similar for all methods, but for the tables side by side: in (d) yearly variations may be noticed in the zonal intensity (higher in the years on the right side, lower in the others), whereas the meridional component seems little varying along years. Only in 2011 a strong increase results of this component that one may put in relation with *La Niña* occurrence, the phenomenon in which significantly colder temperatures than usual result. Note that, unlike in the others, this pattern does not result in the side by side tables analysis [Fig. 1(right(b))].

The pattern of the months, projected as supplemental elements in the principal wind plane, reflects the varying intensity of the trade winds, that increase in the winter months (not shown). The pattern is different in the side by side analysis.

## *4.2 The* CPCA *of Winds*

As already told, the eigenvalues extracted by the complex principal component analysis are all real. In this case, the first principal component is accounted for nearly 29% of the total inertia, the second being accounted for only 12%. Here, we limit our comments to the first complex component. The circle of correlation, shows a pattern of the vectors-buoys, whose length represents the amount of each time series inertia on this first component. On the opposite, the orientation of the eigenvector is arbitrary, thus the buoys position does not reflect zonal and meridional wind components. In Fig. 2(above) the buoys condensed in the first quadrant of the first principal component of *CPCA* appear mostly aligned, with only some buoys situated at the westernmost side of the array and one in the eastern side badly represented.

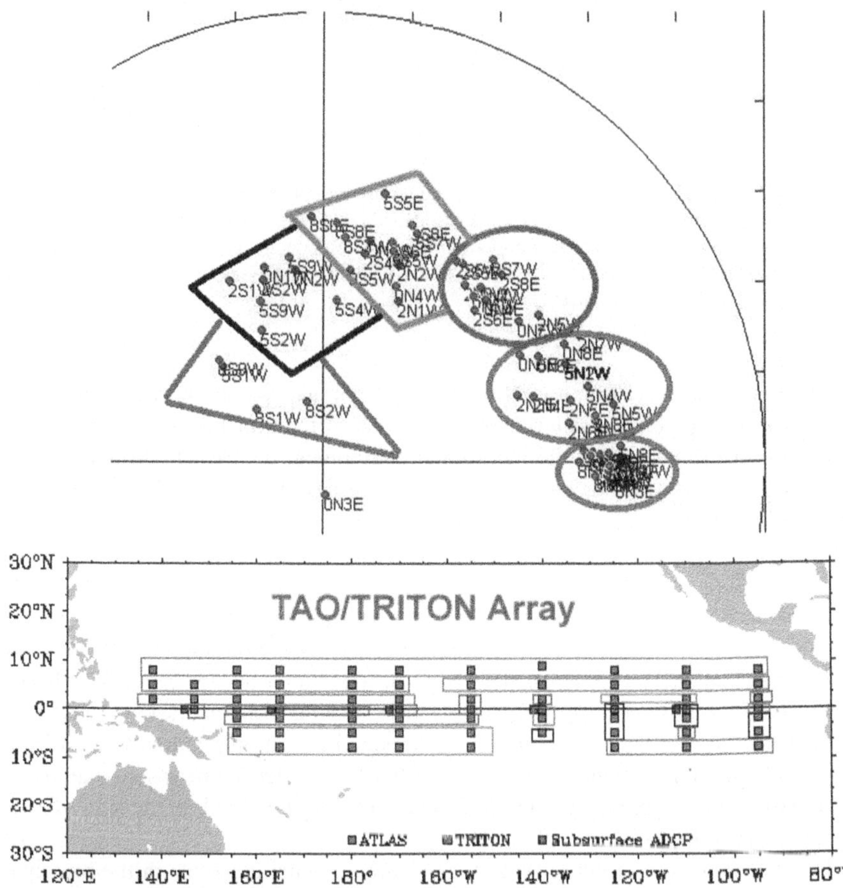


**Fig. 2** Above: representation of the buoys' time series on the plane spanned by the first complex principal component. Below (source NOAA web site): location of the same buoys on the array. Equal contour colors indicate the same group in both images

In Fig. 2(below) the buoys are situated on the array that crosses the Equatorial Pacific Ocean. The leftmost group above corresponds to the extreme south-east buoys, then progressively classes are shifted towards north-west until the rightmost one that corresponds to the northeast and northwestern part of the array. Thus, a rotation of the winds results with respect to the first component, depending on the latitude.

In Fig. 3(above and below) the pattern of the months and of the years is represented in the space of variables, respectively. The first allows us to interpret the principal component, indicating the position of the month according to its average wind; the second does the same for the average wind of each year, allowing the identification of their pattern of variation. Looking at Fig. 3(above) one may recognize the yearly continuous oscillation of the trade winds' intensity, with maxima in September-

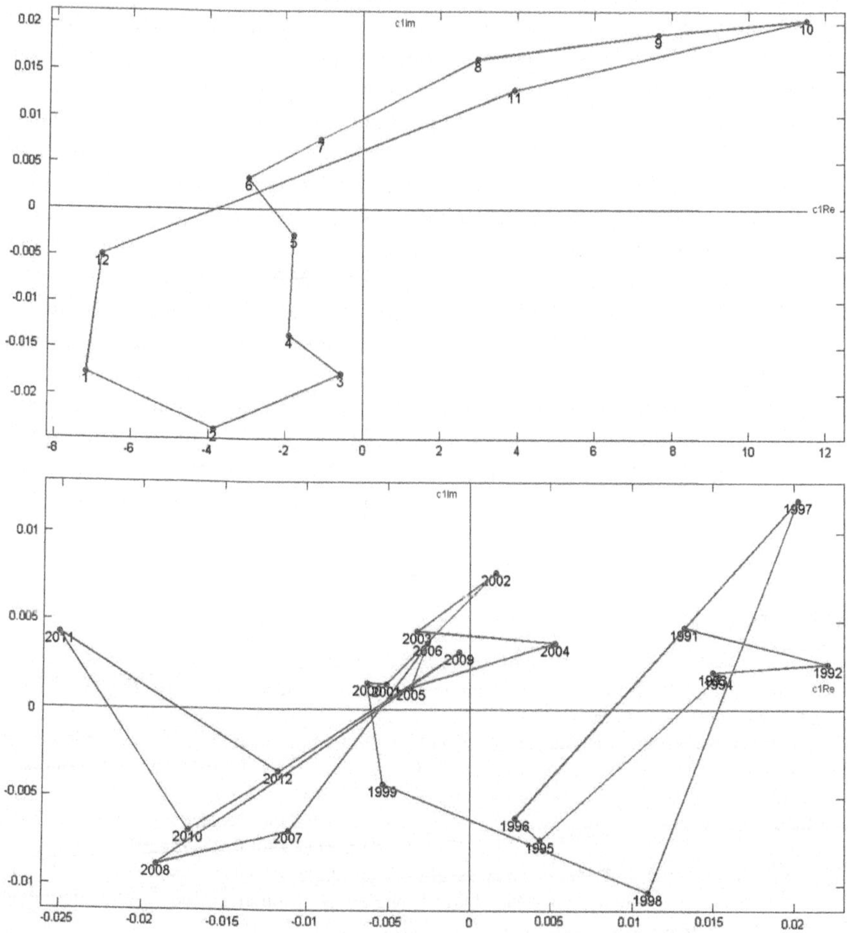


**Fig. 3** Representation of the monthly (above) and of the yearly averages (below) on the plane spanned by the first complex principal component

October and minima in January-February. Note the high scale difference between the two components, indicating a scarce rotation of the winds along the year. The pattern of years in Fig. 3(below) is more complicated and the variation along the two components is comparable. Indeed, some extreme values occur in some years of either *El Niño* or *La Niña* presence: for the first one, consider the position of both 1992 and 1997 at one extreme and for the second both 2008 and 2011 at the other one.

## 5 Conclusion

Summarizing the comparison results, we may say that the first two methods, separate analysis and analysis of the tables side by side, get really difficult the (also physical) interpretation of the relations among components, producing graphics very different from the others. Unlike these, stacked analysis (with splitted components as said) and *CPCA* do not seem to show, at least at first glance, substantial differences between them. Indeed, the *CPCA* rationale seems more adapt to the winds data and its use is simpler than the data manipulations required by the stacked analysis. Concerning *CPCA* of the whole array, the study limited to the first principal component does not seem sufficient to understand the phenomenon, in particular considering that the trade winds variations are not the most evident signs of *El Niño* or *La Niña* presence: this would be rather detected by studying both temperature and pressure in conjunction with winds through a comprehensive exploratory study. In this respect, *CPCA* seems more suitable than the other methods, since it is not manipulated as the others are. Indeed, *CPCA* deserves being theoretically deepened, in order to understand its true meaning in terms of relations between inertia and correlation, something different in real and complex analysis, and to correctly rotate the components, to associate their real and imaginary part to zonal and meridional winds components, respectively.

## References

1. Horel, J.: Complex principal component analysis: theory and examples. J. Clim. Appl. Meteorol. **23**, 1660–1673 (1984)
2. Philander, S.: El Niño, La Niña, and the Southern Oscillation. Academic Press, London (1990)
3. Wang, C., Deser, C., Yu, J.V., Dinezio, P., Clement, A.: El Niño and Southern Oscillation (*ENSO*): A Review. (2012). https://doi.org/10.1.1.364.4359
4. Camiz, S., Denimal, J., Sosa, W.: Exploratory analysis of Pacific Ocean data to study "El Niño" phenomenon. Revista de la Facultad de Ciencias de la *UNI* **13**(1), 50–58 (2010)
5. Camiz, S., Denimal, J., Purini, R.: New results of multidimensional analysis of *TAO/NOAA* aata on "El Niño" phenomenon. In: Hucailuk, C., Núñez, N., Molina, E. (eds.) Actas de trabajos completos *E-ICES*, vol. 9, pp. 24–45. CNEA, Buenos Aires (2014)
6. Jolliffe, I.: Principal Components Analysis. Springer, Berlin (2002)
7. Preisendorfer, R.W.: In: Mobley, C.D. (eds.) Principal Component Analysis in Meteorology and Oceanography. Elsevier, Amsterdam (1988)

8. Bankó, Z., Dobos, L., Abonyi, J.: Dynamic principal component analysis in multivariate time-series segmentation. Conser. Inf. Evol. Sustain. Eng. Econ. **1**(1), 11–24 (2011)
9. Camiz, S., Diblasi, A.: Evolutionary principal component analysis. In: Trabajos Completos, XLI Coloquio Argentino de Estad ística, pp. 680–685. Universidad de Cuyo en Mendoza, Argentina, CD-ROM, 16–18 Octobre 2013
10. Rodrigues, P.C.: Principal component analysis of dependent data. In: 15th European Young Statisticians Meeting September 10–14. Castro Urdiales (Spain) (2007)
11. Autonne, L.: Sur les matrices hypohermitiennes et les unitaires. Comptes rendus des séances hebdomadaires de l' Académie des sciences de Paris **156**, 858–860 (1913)
12. Stewart, G.: On the early history of the singular value decomposition. SIAM Rev. **35**(4), 551–566 (1993)
13. Eckart, C., Young, G.: The approximation of one matrix by another of lower rank. Psychometrika **1**, 211–218 (1936)
14. *NOAA*: Tropical Atmosphere Ocean Project. Pacific Marine Environmental Laboratory (2015)
15. Haykin, S.: Self-organizing Maps in Neural Networks—A Comprehensive Foundation. Prentice-Hall, Upper Saddle River, NJ (1999)
16. *MATLAB*: Release 2012b. The MathWorks, Inc., Natick (MS) (2012)

# Motivations and Expectations of Students' Mobility Abroad: A Mapping Technique

**Valeria Caviezel, Anna Maria Falzoni and Sebastiano Vitali**

**Abstract** Internationalization of higher education is a rapidly rising phenomenon and it has become a priority in the European education policy. At the same time, research in this area is expanding with the aim of understanding motivations and potential benefits of international students' mobility. Within this context, the purpose of our contribution is to analyse students' motivations and the fulfillment of their expectations about the mobility experience abroad. To this aim, we have conducted an online survey addressed to a sample of about 1300 Italian students enrolled in a medium size university (the University of Bergamo) with outward mobility experiences during the six academic years from 2008/2009 to 2013/2014. To assess the results of the survey, we propose a mapping of the answer variables using the VOS*viewer* software.

**Keywords** Credit mobility · Questionnaire · VOS*viewer* · Variables' mapping

## 1 Introduction

In the last decades, an increasing number of European students has spent at least a semester abroad during their university studies. Since it began in 1987/1988, the world's most successful student mobility program, the Erasmus program, has provided over three million European students with the opportunity to go abroad and

V. Caviezel (✉) · A. M. Falzoni
Department of Management, Economics and Quantitative Methods,
University of Bergamo, Via dei Caniana 2, 24127 Bergamo, Italy
e-mail: valeria.caviezel@unibg.it

A. M. Falzoni
e-mail: anna-maria.falzoni@unibg.it

S. Vitali
Department of Probability and Mathematical Statistics, Faculty of Mathematics and Physics,
Charles University, Sokolovska 83, 186 75 Prague, Czech Republic
e-mail: vitali@karlin.mff.cuni.cz

 
F. Mola et al. (eds.), *Classification, (Big) Data Analysis and Statistical Learning*,
Studies in Classification, Data Analysis, and Knowledge Organization,
https://doi.org/10.1007/978-3-319-55708-3_10

study at a higher education institution or train in a company [4]. Even if we should analyze worldwide data on students enrolled outside their country of citizenship, the growth in the internationalization of tertiary education would appear substantial, from 0.8 million in 1975 to 4.5 million students in 2012 [6]. Research in this area is expanding. The studies have mainly focused on the factors influencing the choice to spend a period of study abroad and on the effects that the international mobility can produce on the skills and, eventually, on the employability (see, among others, [1–3, 5, 7, 8]). The purpose of this contribution is to analyze students' motivations and the fulfillment of their expectations. Therefore, we have conducted an online survey involving a sample of about 1300 Italian students enrolled in a medium size university in the North of the country: the University of Bergamo. Our sample is composed of all the credit mobility experiences done by the students during the six academic years from 2008/2009 to 2013/2014. Besides the results of the survey, for each student, we have administrative information collected at enrollment (age, gender, region of residence, etc.) and data on students' academic career. We analyze the students' answers using an approach based on the variables' mapping. Our results show that the students consider their personal development the most important outcome of their international experience. Curiosity about different cultures, the desire to live a new experience and the improvement of foreign language skills are the main factors that motivate the choice of mobility abroad. To improve career prospects and to enhance future employability emerge as motivations (in particular for graduate students), but they are not perceived as the most important outcome of the study abroad experience.

Section 2 describes the questionnaire and the sample, Sect. 3 presents the VOS *viewer* software, Sect. 4 investigates the results of the survey through a mapping of the answers and Sect. 5 offers some conclusions.

## 2 Questionnaire and Sample

To assess the experience of the students of the University of Bergamo involved in this research, we prepared a questionnaire consisting of three sections: *Decision to study abroad* (section B), *Experience abroad* (section C) and *Coming back to Bergamo* (section D). Before these three parts, we ask a few questions (section A) regarding the student's personal details: parents' level of education, parents' employment status, his/her own previous experience abroad or their families', the current employment status of the respondent, the academic year and the number of semesters spent abroad, the type of internationalization program.

In the first section "*Decision to study abroad*", we ask the students to motivate their decision to study abroad, i.e. to enhance future employability, to enrich their CV, to live a new experience, to improve the knowledge of a foreign language, to get in touch with the culture of the host country. Furthermore, we analyze the factors which address the choice towards the host country and the university: i.e. compatibility of study programs and availability of scholarships, prestige of host

**Table 1** Main characteristics of the students involved in the analysis (in %)

| Department | | Gender | | Study | | Semester | | Country | |
|---|---|---|---|---|---|---|---|---|---|
| Engineering | 7.9 | Males | 33 | Bachelor | 66.4 | Fall | 44.6 | Spain | 30.4 |
| Foreign Languages | 51.3 | Females | 67 | Master | 33.6 | Spring | 20.5 | Germany | 16.1 |
| Social Sciences | 5.2 | | | | | Year | 34.9 | France | 15.2 |
| Law | 3.9 | | | | | | | UK | 14.7 |
| Art & Philosophy | 4.1 | | | | | | | Others Euro | 19.0 |
| Economics | 27.6 | | | | | | | Australia | 1.2 |
| | | | | | | | | China | 1.8 |
| | | | | | | | | USA | 1.6 |

city and reputation of the university, knowledge of the language and culture of the host country, living costs.

In the second section "*Experience abroad*", we ask the students to compare the linguistic abilities, the teaching and assessment methods, the average exam evaluation between the experience abroad and the previous experience in Bergamo. Moreover, we investigate the main funding sources and the need to earn a living during the experience abroad.

Finally, in the third section "*Coming back to Bergamo*", the respondents are asked to evaluate the problems in aligning with the home university study program, the time spent studying, the impact of the experience on communication competences, their linguistic and team-working abilities, their adaptability and problem-solving skills. We ask about the fulfillment of their expectations in relation to their personal growth, their linguistic abilities and interpersonal skills, and to the probability to have a better job in the future in Italy or abroad. For each question, there are three/four answer categories, generally measured with an ordinal scale.

In this paper, we focus on the motivations and the satisfaction of the expectations; for this reason only the sections B and D of the questionnaire are analyzed.

Our survey involves 1299 students and former students, which spent one/two semesters abroad for an Erasmus or Extra EU program over the 6 academic years from 2008/2009 to 2013/2014. In Table 1 we summarize the main characteristics of the students: department, gender, bachelor or master, semester spent abroad, country (in %).

## 3 VOS*viewer*

To assess the results of our survey we propose a mapping of the answer variables using the VOS*viewer* software (VOS). VOS (*visualization of similarities*) is a free software created by N. J. van Eck and L. Waltman ten years ago, constantly updated by the authors and downloadable from the website www.vosviewer.com [9, 10, 12]. As an alternative to standard methods for analyzing survey data, the choice to use VOS is given by the easy and immediate interpretability of maps that are obtained. Indeed, the aim of VOS is to provide a two-dimensional visualization in which objects are located in such a way that the distance between any pair of objects reflects their similarity as accurately as possible. Objects that have a high similarity should be located close to each other, whereas objects that have a low similarity should be located far from each other. To map objects the indirect similarities are also considered.

To construct a map, VOS considers a normalization of the co-occurrence matrix. All items ($K = 35$) - relating to the sections B and D of the questionnaire - were dichotomized in such a way that one corresponds to the most positive judgment of the subject (*A lot*) and zero the other categories of answer (*Not at all, A little, Quite a lot*). Therefore the co-occurrence symmetric matrix is constructed in such a way that each element of the main diagonal $c_{ii}$ ($i = 1, 2, \ldots, K$) indicates the number of occurrences of the item $i$, that is the number of students who attributed a very positive judgment to item $i$. The elements of the lower triangular part $c_{ij}$ ($i, j = 1, 2, \ldots, K$ with $i \neq j$) are the number of co-occurrences, that is the number of students who responded with a very positive category both items $i$ and $j$. To obtain a similarity measure [11] between items $i$ and $j$, VOS considers a normalization of the co-occurrence $c_{ij}$ such that $s_{ij} = \frac{c_{ij}}{c_{ii} \cdot c_{jj}}$. The similarity measure $s_{ij}$ is proportional to the ratio between on the one hand the observed number of co-occurrences of items $i$ and $j$ and on the other hand the expected number of co-occurrences of items $i$ and $j$ under the assumption that occurrences of items $i$ and $j$ are statistically independent. If $s_{ij} > 1$, items $i$ and $j$ co-occur more frequently than it would be expected by chance, whereas if $s_{ij} < 1$, items $i$ and $j$ co-occur less frequently than it would be expected by chance.

The idea [10] of VOS mapping technique is to minimize a weighted sum of the squared Euclidian distances between all pairs of items. The higher the similarity between two items, the higher the weight of their squared distance in the summation:

$$\min V(\mathbf{x}_1, \ldots, \mathbf{x}_K) = \sum_{i<j} s_{ij} \left\| \mathbf{x}_i - \mathbf{x}_j \right\|^2$$

where the vector $\mathbf{x}_i = (x_{i1}; x_{i2})$ denotes the location of item $i$ in the two-dimensional map and $\|\bullet\|$ the Euclidian norm. To avoid trivial solutions the constraint that the average distance between two items must be equal to 1 is imposed:

$$\frac{2}{n(n-1)} \sum_{i<j} s_{ij} \left\| \mathbf{x}_i - \mathbf{x}_j \right\| = 1.$$

The constrained optimization problem is solved numerically in two steps (for more details see [10]).

In a VOS map, each variable is indicated by a circle whose dimension depends on the occurrences of the variable, i.e. the number of students choosing the answer at the highest level. The strength of the link between a pair of variables depends on the co-occurrences, i.e. the number of students giving the highest level answer to both variables. Moreover, in a circular map the most important items tend to be located towards the center, the others in the external part.

## 4 Results

After more than one reminder, the answer rate is approximately 50%. The answer rate is increasing: from 43.2% for mobility experiences during the a.y. 2008/2009 to percentages just above 50% for the last two academic years. The answer rate shows a peak for engineering students (64.9%) and a minimum for art and philosophy (42.6%); for the other departments, the number of respondents is around 50%. For this analysis, after the removal of the missing values, we consider 597 questionnaires. In Table 2 we summarize the main questionnaire items and for each one the percentage for each response category.

For a correct interpretation of the maps presented in this section, we remind that labels beginning with B refer to the first section of the questionnaire - *Decision to study abroad* - while the labels beginning with D refer to the third section - *Coming back to Bergamo*.

The map in Fig. 1, based on all 597 respondents, highlights that students want to participate in a mobility program abroad mainly to satisfy the curiosity and the desire to live a new experience (B06 Curiosity), to improve their knowledge of a foreign language (B06 Improve Language), to enrich the CV (B06 Improve CV), and to know the culture of the host country (B06 Host Country). They select the destination principally taking into account the language and the culture of the potential host country (B08 Cultures Country) and the appeal of the host city (B08 Appeal City). Our results show that, when back to the home country, the respondents judge their expectations fulfilled. Indeed, they evaluate that the experience of studying abroad has had a very positive impact on their adaptability (D05 Adaptability), linguistic abilities (D05 Linguistic Ability), communication (D05 Communication Skills) and problem-solving skills (D05 Problem- Solving). The satisfaction in terms of personal growth (D06 Personal Growth), linguistic abilities (D06 Linguistic Ability) and interpersonal skills (D06 Interpersonal skills) is remarkable. In the map of Fig. 1, the relationship between the motivations and the most important goals expected from the mobility experience is highlighted by the thirty strongest links which show the eight items most connected to each other. Among these eight variables, the main motivations are curiosity and language improvement, while the other six variables represent the fulfillment of expectations in terms of development of soft skills and foreign language's improvement. These results are aligned with the findings of the survey carried out in [1] at the European level. Quite surprisingly, to enhance the future employability is neither among the most important motivations (B06 Future Work), nor among the main expectations satisfied (D06 IT Job Market and D06 Abroad Job Market). A possible reason of these answers is that respondents are mainly students and, at the time of the survey, they lack of a direct experience in the job market.

**Table 2** Main questionnaire items and corresponding percentage of responses

| Description | Label | Not at all | A little | Quite a lot | A lot |
|---|---|---|---|---|---|
| How important were the following factors in motivating your decision to study abroad? | B06 Curiosity | 0.2 | 1.0 | 10.2 | 88.6 |
| | B06 Improve Language | 0.3 | 1.0 | 10.9 | 87.8 |
| | B06 Improve CV | 1.5 | 8.7 | 41.2 | 48.6 |
| | B06 Host Country | 1.8 | 11.1 | 39.0 | 48.1 |
| | B06 Future work | 2.7 | 20.4 | 41.9 | 35.0 |
| How important were the following factors in choosing the destination of your study abroad? | B08 Culture country | 9.4 | 15.2 | 33.5 | 41.9 |
| | B08 Appeal city | 5.2 | 24.3 | 39.4 | 31.1 |
| Did you have any problem in sitting any exam not taken prior to departure? | D02 Sitting exam | 6.0 | 23.3 | 37.2 | 33.5 |
| Did you have any problem in aligning your progress in university? | D03 Aligning study | 4.9 | 22.1 | 34.1 | 38.9 |
| How important was the impact of your experience abroad on the following factors? | D05 Adaptability | 0.2 | 1.5 | 20.3 | 78.0 |
| | D05 Linguistic ability | 0.7 | 0.8 | 26.5 | 72.0 |
| | D05 Communication skills | 0.2 | 3.7 | 33.5 | 62.6 |
| | D05 Problem-solving | 1.3 | 5.5 | 41.4 | 51.8 |
| How satisfied were your expectations about the following factors? | D06 Personal growth | 0.0 | 0.3 | 11.6 | 88.1 |
| | D06 Linguistic ability | 0.3 | 1.8 | 22.5 | 75.4 |
| | D06 Interpersonal skills | 0.2 | 3.3 | 26.8 | 69.7 |
| | D06 IT job market | 6.7 | 29.3 | 44.6 | 19.4 |
| | D06 Abroad job market | 2.0 | 12.9 | 53.6 | 31.5 |

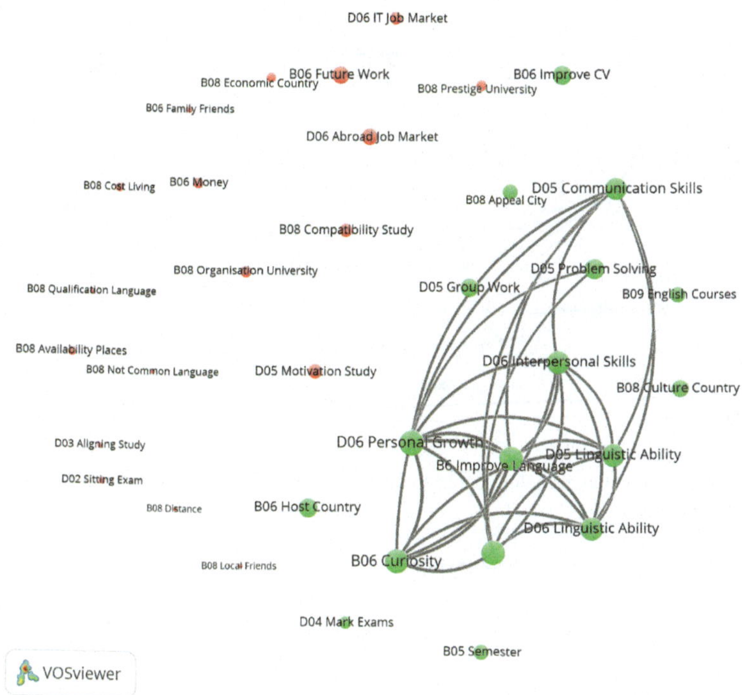


**Fig. 1** Map based on the whole sample dataset

Further evidence on this point emerges from Figs. 2 and 3 which show the map of bachelor and master degree students, respectively. Both of them are a focus of the items highlighted by the first thirty links. Not surprisingly, the difference emerging between the two groups of students concerns the role of mobility abroad on employability and career prospects. The improvement of the CV (B06 Improve CV) is a very important factor which motivates the choice to study abroad for the 54% of the master degree students compared to the 46% of the bachelor degree ones. However, for both groups of students, the improvement of career opportunities in Italy and abroad does not emerge as the most important expected outcome of the international mobility experience. These findings cannot be interpreted as evidence of a lack of impact on employability of the study abroad experience: different data would be required. As discussed in studies [3, 7, 8], to test the causal effect of international mobility on employability requires to compare, after graduation, mobile with nonmobile students, and to control for pre-mobility differences between students.

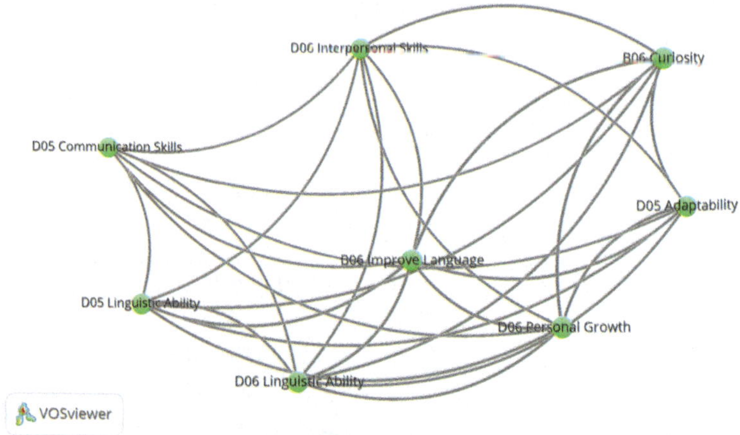


**Fig. 2** Map based on the bachelor degree students' dataset - Focus on the items connected by the first thirty links

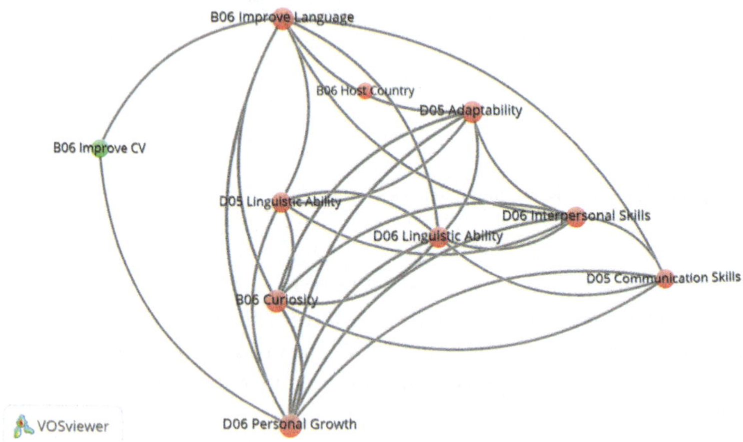


**Fig. 3** Map based on the master degree students' dataset - Focus on the items connected by the first thirty links

## 5 Conclusions

Internationalization of higher education is a rapidly rising phenomenon and it has become a priority in the European education policy. One of the 2020 EU targets is that at least 20% of higher education students graduates with a period abroad of either higher education-related study or training. Research in this area may contribute to understand the potential advantages associated with participation in study abroad programs and may provide rationales for countries and for public and private institutions to facilitate the international student mobility. The rising number of applications show that students are aware of the importance of this experience. However, barriers still exist such as financial constraints, problems with compatibility of courses and credit recognition, lack of information and support at the home institution. Actions taken by universities to overcome these obstacles may act as incentives in extending the group of internationally mobile students.

The purpose of this contribution has been to analyze students' motivations and the fulfillment of expectations of their mobility experience abroad. Our results show that studying abroad is motivated first of all by curiosity and the desire to live a new experience, together with the aim to improve a foreign language. In terms of achieved objectives, the students put at the top the personal growth and the development of communication and interpersonal skills. To enhance employability in Italy and abroad emerges as a motivation (in particular for graduate students), but not perceived yet as an important effect of the study abroad experience.

## References

1. CHE Consult, Brussels Education Services, Centrum fur Hochschulentwicklung, Compostela Group of Universities, Erasmus Student Network: Erasmus impact study. Effects of mobility on the skills and employability of students and the internationalization of higher education institutions. European Commission, Brussels (2014). https://doi.org/10.2766/75468
2. Di Pietro, G., Page, L.: Who studies abroad? Evidence from France and Italy. Eur. J. Educ. **43**, 389–398 (2008)
3. Di Pietro, G.: Do study abroad programs enhance the employability of graduates? Educ. Finan. Policy **10**(2), 223–243 (2015)
4. European Commission: Erasmus—Facts, Figures & Trends. Publications Office of the European Union, Luxembourg (2015). http://ec.europa.eu/dgs/education_culture
5. Luo, J., Jamieson-Drake, D.: Predictors of study abroad intent, participation, and college outcomes. Res. High. Educ. **56**, 29–56 (2015)
6. OECD: OECD Education at a Glance 2015, Paris (2015). https://doi.org/10.1787/eag-2017-en
7. Parey, M., Waldinger, F.: Studying abroad and the effect on international labour market mobility: evidence from the introduction of ERASMUS. Econ. J. **121**, 194–222 (2010)
8. Rodrigues, M.: Does student mobility during higher education pay? Evidence from 16 European Countries. JRC Scientific and Policy Reports (2013). https://doi.org/10.2788/95642
9. Van Eck, N. J., Waltman, L.: A new method for visualizing similarities between objects. In: Lenz, H.J., Decker, R. (eds.) Advances in Data Analysis: Proceedings of the 30th Annual Conference of the German Classification Society, pp. 299–306. Springer, Berlin (2007)
10. Van Eck, N.J., Waltman, L.: Software survey: VOSviewer, a computer program for bibliometric mapping. Scientometrics **84**, 523–538 (2010)
11. Van Eck, N.J., Waltman, L.: How to normalize co-occurrence data? An analysis of some well-known similarity measures. J. Am. Soc. Inf. Sci. Technol. **60**(8), 1635–1651 (2009)
12. Van Eck, N.J., Waltman, L.: Visualizing bibliometric networks. In: Ding, Y., Rousseau, R., Wolfram, D. (eds.) Measuring Scholarly Impact: Methods and Practice, pp. 285–320. Springer, Berlin (2014)

# Testing Circular Antipodal Symmetry Through Data Depths

**Giuseppe Pandolfo, Giovanni Casale and Giovanni C. Porzio**

**Abstract** This work discusses how to test antipodal symmetry of circular distributions through depth functions. Two notions of depths for circular data are adopted, and their performances are evaluated and compared through a simulation study.

**Keywords** Angular data depth · Nonparametric statistics · $\ell$-fold symmetric distributions

## 1 Introduction

Circular statistics deals with data that can be represented as points on the circumference of the unit circle. Data of this type are themselves referred to as being circular, a term used to distinguish them from linear data (i.e., data lying on a line).

Circular data arise from a variety of sources. Many examples of circular data can be found in various scientific fields such as earth sciences, meteorology, biology, physics, psychology, and medicine. Some common examples are the migration paths of birds and animals, wind directions, ocean current directions, and patients' arrival times in an emergency ward of a hospital. Circular data also include directions measured using instruments such as a compass, protractor, weather vane, sextant or theodolite. Such directions are recorded as angles expressed in degrees or radians, measured either clockwise or counter-clockwise from some origin, referred to as the zero direction. Each angle defines a point on the circumference of the unit circle, just as each value of a linear variable defines a point on the real line. What is specific

G. Pandolfo (✉)
Department of Industrial Engineering, University of Naples Federico II, Naples, Italy
e-mail: giuseppe.pandolfo@unina.it

G. Casale · G. C. Porzio
Department of Economics and Law, University of Cassino and Southern Lazio, Cassino, Italy
e-mail: g.casale@unicas.it

G. C. Porzio
e-mail: porzio@unicas.it



F. Mola et al. (eds.), *Classification, (Big) Data Analysis and Statistical Learning*, Studies in Classification, Data Analysis, and Knowledge Organization,
https://doi.org/10.1007/978-3-319-55708-3_11

to circular data is their periodic nature, that in turns forces to abandon standard statistical techniques designed for linear data.

Along with the increasing availability of circular data, the growing attention on asymmetric circular models (see, e.g., [2, 11]) suggests that testing circular symmetries is a topic of interest. For this reason, this work aims at evaluating procedures to test if a circular distribution is antipodal symmetric.

Within the next section, a brief introduction to circular symmetries is first offered. Then, two notions of depth functions for angular data are presented, and the test for antipodal symmetry on the circle suggested in [6] is described. Furthermore, as a novelty, a test using the arc distance depth is introduced. Finally, a comparison of the performances of the two test procedures is offered by means of a simulation study.

## 2 Reflective and Antipodal Symmetry

An important question in statistics is whether a distribution is symmetric, and this is true also for circular data. However, symmetry of circular data deserves a special attention, given that different types of circular symmetry can be defined. Specifically, a distribution which is symmetric about an axis is defined reflectively symmetric, while it is called $\ell$-fold symmetric if applying a $2\pi/\ell$ rotation the shape of the distribution does not change.

Formally, a circular continuous distribution $H$ with density $h(\cdot)$ is defined to be reflectively symmetric about a central direction $\phi$ if $h(\theta) = h(2\phi - \theta)$ (see [8]). Reflective symmetry substantially corresponds to the standard symmetry adopted for linear data. However, while on the line testing for symmetry about a measure of location gave birth to a great number of publications, the corresponding literature seems to be sparse for circular data. A few testing procedures are available to test the null hypothesis of circular reflective symmetry. A locally rank test that comprises the circular sign and Wilcoxon tests was proposed in [10]. Pewsey [8] in 2002 introduced a simple omnibus test for reflective symmetry based on the sample second sine moment about an estimation of an unknown central direction. An adaptation of the runs tests in [7] to the circular setting can be found in [9]. More recently, optimal tests for circular reflective symmetry about a fixed median direction were proposed in [5].

As for the *$\ell$-fold* symmetry, it is formally defined as follows. A circular continuous distribution $H$ with density $h(\cdot)$ is *$\ell$-fold* symmetric if $h(\theta) = h(\theta + 2\pi/\ell)$. For $\ell = 2$, a special kind of symmetry (called antipodal symmetry) is obtained. If continuous, an antipodal symmetric distribution has $h(\theta) = h(-\theta)$ for any direction $\theta$ on the circle.

Within the literature, we found two rank-based tests to assess the null hypothesis of *$\ell$-fold* symmetry (that can be obviously used for $\ell = 2$; see [3]), and some clues to test antipodal symmetry in [6]. They suggested a test can be obtained by exploiting a characterization of a depth function in case of antipodality, as discussed in the next section.

## 3 Data Depth-Based Tests for Antipodal Symmetry

Data depth functions on circles and spheres lead to proper notions of center (or depth-based median) and an inner-outward ranking of directional data with respect to the center of the distribution.

Four notions of data depth for directional data are available: the angular simplicial depth, the angular Tukey's depth, the arc distance depth (see [6]), and the angular Mahalanobis depth (see [4]). The first two extend the notions of simplicial and Tukey's depths in $\mathbb{R}^d$ to circles and spheres. The notion of arc distance depth is instead inspired by the $L^1$ distance in the Euclidean space, while the angular Mahalanobis depth is a canonical notion of depth for rotationally symmetric distributions.

The constancy of the first three depth functions has been fully studied in [6], and they found that the constancy of the simplicial and of the arc distance depths characterize antipodal symmetry. This allows defining depth-based tests for antipodal symmetry.

### 3.1 The Angular Simplicial and the Arc Distance Depths

Formally, the angular simplicial depth (ASD) of the point $\theta$ w.r.t. the distribution $H$ on the circle is defined as follows:

$$ASD\,(\theta, H) := P_H\,(\theta \in arc\,(W_1, W_2))$$

where $P_H$ denotes the probability content w.r.t. the distribution $H$, $W_1$ and $W_2$ are i.i.d. observations from $H$, and $\mathrm{arc}(w_1, w_2)$ is the shortest arc joining the points $w_1$ and $w_2$.

The arc distance depth (ADD) of the point $\theta$ w.r.t. the distribution $H$ on the circle is defined as follows :

$$ADD\,(\theta, H) := \pi - \int L\,(\theta, \varphi)\,dH\,(\varphi)$$

where $L\,(\theta, \varphi)$ is the Riemannian distance between $\theta$ and $\varphi$ (i.e. the length of the shortest arc joining $\theta$ and $\varphi$).

As noted in [6] (p. 1484), the angular simplicial depth "provides a finer ranking of data points in the order of centrality". On the other hand, it should be mentioned that the arc distance depth has the significant advantage to require a substantially lower computational effort to be computed.

For these two functions, the following theorems are proved in [6]:

**Theorem 1** (Constancy of the angular simplicial depth on the circle) *Let $H$ be a circular continuous distribution, with density $h\,(\cdot)$. Then $ASD\,(\theta, H) = c$, for a pos-*

*itive constant c and all* $\theta \in [0, 2\pi)$*, if and only if* $h(\theta) = h(-\theta)$ *for all* $\theta$*. Moreover, the constant c must then be equal to* $1/4$.

**Theorem 2** (Constancy of the arc distance depth on the circle) *Let* $H$ *be a circular continuous distribution, with density* $h(\cdot)$*. Then* $ADD(\theta, H) = c$*, for a positive constant c and all* $\theta \in [0, 2\pi)$*, if and only if* $h(\theta) = h(-\theta)$ *for all* $\theta$*. Moreover, the constant c must then be equal to* $\pi/2$.

In brief, constancy of both these two depth functions characterizes antipodal symmetric distributions.

## 3.2 Tests for Antipodal Symmetry

Exploiting the results reported above on the characterization of the angular simplicial depth, a possible test for antipodal symmetry may use the following test statistic (see [6]):

$$T_n^{AS} := \sup_{\theta} |ASD(\theta, H_n) - c_{ASD}| \tag{1}$$

where $ASD(\theta, H_n)$ is the value attained by the empirical angular simplicial depth at the point $\theta$ and $c_{ASD}$ is the constant characterizing the simplicial depth in case of antipodal symmetry (1/4 for circular distributions). Large values of the test statistic suggest that the antipodal symmetry hypothesis is unlikely to be true.

Given that neither the exact nor an approximate sampling distribution of this test statistic are known, a resampling method in order to evaluate how much the data support/do not support the null hypothesis of antipodal symmetry was suggested in [6].

After them, given an observed sample of size $n$, the observed significance level for the test in (1) is obtained as follows:

1. compute the test statistic for the observed sample;
2. draw with replacement a sample of size $n$ from the given observed sample;
3. apply a rotation of $\pi$ to a subset of $n/2$ observations randomly selected from the sample drawn at step 2;
4. compute the test statistic on the modified sample obtained at step 3;
5. repeat steps 2, 3 and 4, $R$ times;
6. estimate the observed significance level by computing the proportion of test statistic values obtained at step 4 greater or equal than the value of the test statistic obtained at step 1.

This resampling procedure differs from bootstrap in step 3, where a random reflection is applied to the original data in order to obtain new samples of size $n$ on which the test statistic values are computed. These values are used to approximate the sampling distribution under the null hypothesis of antipodal symmetry. This method was adopted in [1] to evaluate the presence of a preferred direction of some orientation data.

However, it is worth underlining that the computational complexity of the angular simplicial depth may represent a drawback for this procedure, especially in case of large sample sizes (the computational complexity is $O(n^{d+1})$ in $d$ dimensions).

This in turn motivates an alternative test procedure based on the arc distance depth, which is a function with a lower computational complexity. Hence, in analogy with the test statistic in (1), the following can be considered:

$$T_n^{AD} := \sup_{\theta} |ADD(\theta, H_n) - c_{ADD}| \qquad (2)$$

where $ADD(\theta, H_n)$ is the value attained by the empirical arc distance depth at the point $\theta$ and $c_{ADD}$ is the constant characterizing the arc distance depth in case of antipodal symmetry ($\pi/2$ for circular distributions).

Large values of this test statistic also suggest that the antipodal symmetry hypothesis is unlikely to be true, and the same resampling procedure described above can be adopted to evaluate the observed significance level.

## 4 Evaluating the Test Procedures: An Empirical Study

To the best of our knowledge, no studies are available on the observed test level attained by the test procedure suggested in [6] nor on its power under some alternatives. For this reason, a simulation study was designed with the additional aim of comparing observed levels and power functions with those attained by the test statistic in (2). Specifically, we evaluated whether the nominal significance level is maintained under the null hypothesis, and which of the tests is more powerful against some alternative hypotheses (for two different sample sizes).

### *4.1 Simulation Design*

Data were generated from an equal-weighted mixture of von Mises distributions $vM(\mu, \kappa)$ with means at $\mu_1$ and $\mu_2$, and equal concentration parameters (the dispersion parameters) $\kappa_1 = \kappa_2 = 10$. This mixture model is formally expressed as follows:

$$Y = (1 - \lambda) X_1 + \lambda X_2,$$

where $X_1 \sim vM(\mu_1, 10)$, $X_2 \sim vM(\mu_2, 10)$, and with $\lambda = 0.5$ and $\mu_1$ kept fixed at $0°$.

With $\mu_2 = 180°$, data are generated under the null hypothesis of antipodal symmetry. Conversely, under the alternative hypothesis $\mu_2$ takes values far away from $180°$. We set:

$$\mu_2 = \{180°, 175°, 170°, 165°, 155°, 145°, 140°, 130°, 120°, 90°\}.$$

The further $\mu_2$ is from 180°, the further the data are from the null hypothesis. After [5], simulations were run by generating samples of sizes 30 and 100. A total of $2 \times 10$ simulation conditions were thus considered, and for each of them the observed significance level was computed $N = 1000$ times, according to the resampling procedure described in Sect. 3.2 and by setting $R = 999$ at step 5.

### 4.2 Simulation Results: Nominal Versus Observed Significance Level

Resampling procedures do not imply that the probability to exceed the critical value is equal to the expected one. That is, the *observed level* of the test is not necessarily equal to the *nominal level*. Commonly, scholars tend to tolerate an observed level that is lower than the nominal. In such a case, a test is called *conservative*.

Table 1 reports how many times in $N = 1000$ the observed level is larger than the nominal one under the null hypothesis (that is, with $\mu_2 = 180°$), and $n = 100$. Three nominal test levels were investigated, that is $\alpha = \{0.01, 0.05, 0.10\}$.

Nominal and observed levels do not differ too much for every level of $\alpha$. However, both tests seems to be not conservative, with the simplicial-based test performing slightly worse than the arc distance-based test (being this latter conservative for $\alpha = 0.01$).

### 4.3 Simulation Results: Power of the Tests

Power curves were examined for tests of level $\alpha = \{0.01, 0.05, 0.10\}$. Results for $\alpha = \{0.01, 0.05\}$ are reported here (Fig. 1), given that no qualitative difference arose for the case of $\alpha = 0.10$. The four curves in each panel of Fig. 1 refer to the power of the test based on the simplicial depth for $n = 30(\cdot - \cdot)$, and $n = 100(\text{- - -})$; for the test based on the arc distance depth for $n = 30(\cdots)$, and $n = 100(\text{- - -})$.

First of all, it should be noted that all the power functions are increasing, and the functions for $n = 100$ dominate those obtained for $n = 30$. Furthermore, in the

**Table 1** Number of times in $N = 1000$ the observed significance level of the tests is larger than the nominal one under the null hypothesis (that is, with $\mu_2 = 180°$). Sample size $n = 100$

| Nominal significance level | Observed significance level | |
|---|---|---|
| | Simplicial-based test | Arc distance-based test |
| 0.01 | 16/1000 | 7/1000 |
| 0.05 | 63/1000 | 59/1000 |
| 0.10 | 116/1000 | 109/1000 |

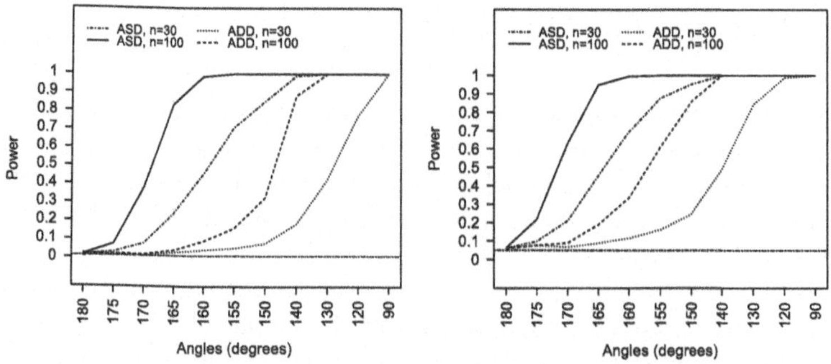


**Fig. 1** Estimated power curves of the tests for antipodal symmetry based on the simplicial and arc distance depth functions, for sample sizes $n = 30$ and $n = 100$. Results are shown for two test levels ($\alpha = 0.01$ panel (a); $\alpha = 0.05$, panel (b)), for the angular simplicial (· — ·—, $n = 30$; ——, $n = 100$), and the arc distance (······ , $n = 30$; - - - -, $n = 100$). The horizontal dot-dashed line indicates the test level

comparison of the power attained by the test statistic in (1) and in (2), we have that the test based on the simplicial depth performs much better. This result holds for each test level, for each sample size, and for each examined value of $\mu_2$ (i.e. under different alternative hypotheses).

To exemplify, for $\alpha = 0.01$, and $n = 100$, when the power of the simplicial-based test exceeds 0.9, the corresponding power of the arc distance-based test is still lower than 0.05. The same kind of effect occurs for $\alpha = 0.05$ and for any sample size.

## 4.4 Simulation Results: Computational Costs

We timed the computation of both methods for $n = 30$ and $n = 100$. We implemented the methods in R and simulations were run on a 2.60 GHz CPU. The average execution times in seconds over 1000 trials are reported in Table 2. It can be seen that computing the arc distance-based test is about 3 times faster on average than the simplicial-based test.

**Table 2** Average execution times in seconds (over 1000 trials) of the tests, for sample sizes $n = 30$ and $n = 100$

| Sample size | Simplicial-based test | Arc distance-based test |
|---|---|---|
| 30 | 4.967 | 1.580 |
| 100 | 10.986 | 3.568 |

## 5 Findings and Final Remarks

The empirical behavior of a test of antipodal symmetry of circular data introduced in [6] was investigated throughout a simulation study. To summarize, results showed that the observed significance level is not far from the nominal level, and that the power function seems to perform quite well under some alternatives, for different test levels and sample sizes.

Furthermore, a new test based on the arc distance depth was introduced, and its performance was analysed in comparison with the simplicial-based test. This latter clearly outperforms the former in terms of power. However, simulation results showed that the arc distance-based test is more conservative and requires less computational cost, since its computational time was about three times shorter than that required by the simplicial-based test.

What is left to further work is the evaluation of these procedures under different alternative hypotheses and the extension of the test to spherical or hyper-spherical data. In these cases, the trade-off between power and computational cost may lead to prefer the arc distance-based test.

## References


1. Agostinelli, C., Romanazzi, M.: Nonparametric analysis of directional data based on data depth. Environ. Ecol. Stat. **20**, 253–270 (2013)
2. Jones, M.C., Pewsey, A.: Inverse Batschelet distributions for circular data. Biometrics **68**, 183–193 (2012)
3. Jupp, P.E., Spurr, B.D.: Sobolev tests for symmetry of directional data. Ann. Stat. **11**, 1225–1231 (1983)
4. Ley, C., Sabbah, C., Verdebout, T.: A new concept of quantiles for directional data and the angular Mahalanobis depth. Electr. J. Stat. **8**, 795–816 (2014)
5. Ley, C., Verdebout, T.: Simple optimal tests for circular reflective symmetry about a specified median direction. Stat. Sin. **24**, 1319–1339 (2014)
6. Liu, R., Singh, K.: Ordering directional data: concepts of data depth on circles and spheres. Ann. Stat. **20**, 1468–1484 (1992)
7. Moddares, R., Gastwirth, J.L.: A modified runs test for symmetry. Stat. Probab. Lett. **31**, 107–112 (1996)
8. Pewsey, A.: Testing circular symmetry. Can. J. Stat. **30**, 591–600 (2002)
9. Pewsey, A.: Testing for circular reflective symmetry about a known median axis. J. Appl. Stat. **31**, 575–585 (2004)
10. Schach, S.: Nonparametric symmetry tests for circular distributions. Biometrika **56**, 571–577 (1969)
11. Umbach, D., Jammalamadaka, S.R.: Building asymmetry into circular distributions. Stat. Probab. Lett. **79**, 659–663 (2009)

# Part IV
# Statistical Modeling

# Multivariate Stochastic Downscaling for Semicontinuous Data

**Lucia Paci, Carlo Trivisano and Daniela Cocchi**

**Abstract** The paper proposes a Bayesian hierarchical model to scale down and adjust deterministic weather model output of temperature and precipitation with meteorological observations, extending the existing literature along different directions. These non-independent data are used jointly into a stochastic calibration model that accounts for the uncertainty in the numerical model. Dependence between temperature and precipitation is introduced through spatial latent processes, at both point and grid cell resolution. Occurrence and accumulation of precipitation are considered through a two-stage spatial model due to the large number of zero measurements and the right-skewness of the distribution of positive rainfall amounts. The model is applied to data coming from the Emilia-Romagna region (Italy).

**Keywords** Weather numerical forecasts · Temperature · Precipitation
Hierarchical modeling · BICAR prior

## 1 Introduction

Numerical models are widely used to estimate and forecast several environmental variables [2]. Their output is provided in terms of averages over grid cells, usually at high spatial and temporal resolution. However, these outputs are often biased and the amount of potential calibration is unknown. Also, numerical models do not associate any measure of uncertainty with their numerical output, being derived from deterministic specifications.

---

L. Paci (✉) · C. Trivisano · D. Cocchi
Department of Statistical Sciences, Università Cattolica del Sacro Cuore,
Largo Gemelli 1, Milan, Italy
e-mail: llucia.paci@unicatt.it

C. Trivisano
e-mail: carlo.trivisano@unibo.it

D. Cocchi
e-mail: daniela.cocchi@unibo.it



F. Mola et al. (eds.), *Classification, (Big) Data Analysis and Statistical Learning*,
Studies in Classification, Data Analysis, and Knowledge Organization,
https://doi.org/10.1007/978-3-319-55708-3_12

Calibration and uncertainty quantification of numerical model output are important statistical goals that are tackled via data fusion, i.e., statistical models that combine deterministic output with data collected at monitoring stations which provide the true levels of the variables. The spatial misalignment between point-referenced monitoring data and gridded numerical model output can be addressed using a downscaling approach, i.e., moving from the lower resolution (grid-level) to the higher resolution (point-level). The univariate downscaler to fuse monitoring data with numerical model output has been introduced by [3] with neighbor-based extensions proposed by [4] to smooth the entire air quality model output and account for its uncertainty.

The downscaling literature is usually based on Gaussian processes. However, the Gaussian assumption is not suitable when dealing with semicontinuous data, that often arise in environmental problems, such as precipitation forecasts. Semicontinuous data are customarily analyzed using two-part models [8] consisting of a binary component that specifies the probability of a positive response (i.e. precipitation occurrence) and a continuous component that models the positive response (i.e. precipitation accumulation).

Most works on downscaling focus on univariate problems, while numerical models usually provide joint estimates of several environmental variables. Here, we consider joint downscaling of temperature and precipitation output arising from a weather numerical model over a common spatial domain. The physical relationship between rainfall and temperature is usually described by a numerical model through a set of deterministic equations. For this reason, bias-correction procedures of numerical models applied to temperature and precipitation separately do not preserve the dynamical link specified by the computer model. Rather, the information regarding one variable will help to improve calibration and prediction of the other.

We propose a Bayesian hierarchical approach to scale down and jointly calibrate temperature and precipitation output from a numerical model, offering a multivariate extension of the neighbor-based downscaler of [4]. We stress the characterization of uncertainty in the numerical model output via a stochastic modeling to allow the propagation of such uncertainty from the deterministic output to the inference on the responses. For univariate temperature, we rely on the Gaussian assumption while, for univariate precipitation, we employ a two-stage mixture model to accommodate point masses at zero and model positive precipitation accumulation at both spatial resolutions.

## 2 Joint Spatial Modeling

Let $Y_1(\mathbf{s})$ and $Y_2(\mathbf{s})$ denote, respectively, the observed temperature and precipitation at site $\mathbf{s}$. Also, let $X_1(B)$ and $X_2(B)$ be the numerical model output for temperature and precipitation at grid cell $B$, respectively. Let $B_s$ be the grid cell that contains the site $\mathbf{s}$; then, the model for the temperature is

$$Y_1(\mathbf{s}) = \beta_0 + \beta_1 \widetilde{X}_1(B_s) + \zeta_1(\mathbf{s}) + \varepsilon_1(\mathbf{s}) \tag{1}$$

where $\zeta_1(\mathbf{s})$ are spatially-structured random effects and $\varepsilon_1(\mathbf{s})$ are pure errors from a white noise process with nugget variance $\tau_1^2$, i.e., $\varepsilon_1(\mathbf{s}) \sim N(0, \tau_1^2)$. In (1), $\widetilde{X}_1(B)$ represents a grid-level latent process driving the temperature output $X_1(B)$, that is

$$X_1(B) = \mu_1 + \widetilde{X}_1(B) + \eta_1(B) \tag{2}$$

where $\mu_1$ is a global mean and $\eta_1(B)$ is distributed as $N(0, \lambda_1^2)$. Details on $\widetilde{X}_1(B)$ are deferred to the second stage of modeling.

As mentioned in the Introduction, a special modeling effort is required for precipitation occurrence and accumulation. Indeed, challenges arise both from the large number of zero measurements (no precipitation) and the right-skewness of the distribution of positive rainfall amounts. In other words, we need to account for point mass at zero as well as to model the precipitation accumulation by continuous distributions. To accomplish that, we adopt a two-stage hierarchical model for precipitation, $Y_2(\mathbf{s})$, that is specified as mixture of a point mass at zero and a lognormal density for the continuous distribution of precipitation amounts conditional on rainfall occurrence. Other models can be chosen for the nonzero precipitation accumulation, such as the Gamma density [5].

To facilitate the computation, we employ an augmented approach [1] that reformulates the binary component of the precipitation occurrence into a continuous problem by introducing a point-level latent Gaussian process, $Y_2^*(\mathbf{s})$, that regulates precipitation occurrence such that there is no precipitation if $Y_2^*(\mathbf{s}) \leq 0$, i.e.

$$Y_2(\mathbf{s}) = \begin{cases} \exp(U_2(\mathbf{s})) \text{ if } & Y_2^*(\mathbf{s}) > 0 \\ 0 & \text{otherwise} \end{cases} \tag{3}$$

First, we model the precipitation occurrence by specifying the latent process $Y_2^*(\mathbf{s})$

$$Y_2^*(\mathbf{s}) = \alpha_0 + \alpha_1 \widetilde{X}_2(B_s) + \varepsilon_2^*(\mathbf{s}) \tag{4}$$

where $\widetilde{X}_2(B_s)$ comes from a grid-level latent process described in the second stage of modeling and $\varepsilon_2^*(\mathbf{s}) \sim N(0, 1)$; the unit variance of $\varepsilon_2^*(\mathbf{s})$ ensures the identifiability of the model. Equation (4) is equivalent to model the precipitation probability via a probit link where the probability of observed precipitation is modeled as a linear function of the underlying rainfall process $\widetilde{X}_2(B_s)$, referred to the grid cell containing site $\mathbf{s}$.

We now turn to the precipitation model output, $X_2(B_s)$. Again, we need to account for point masses at zero (no precipitation output occurrence) and model the precipitation accumulation output by a distribution on a positive domain. Hence, similarly to (3), we assume that $X_2(B)$, is positive when an areal-level latent process, denoted by $X_2^*(B)$ is positive, that is

$$X_2(B) = \begin{cases} \exp(P_2(B)) \text{ if } & X_2^*(B) > 0 \\ 0 & \text{otherwise} \end{cases} \tag{5}$$

Then, analogously to (4), we specify the precipitation occurrence simulated by the numerical model as a linear function of the random effects $\widetilde{X}_2(B)$, that is

$$X_2^*(B) = \mu_2^* + \widetilde{X}_2(B) + \eta_2^*(B) \tag{6}$$

where $\mu_2^*$ is an overall mean and $\eta_2^*(B) \sim N(0, 1)$.

The second part of our model regards precipitation accumulation. We assume a spatial downscaler model for the observed log-precipitation, that is

$$U_2(\mathbf{s}) = \gamma_0 + \gamma_1 \widetilde{X}_2(B_s) + \zeta_2(\mathbf{s}) + \varepsilon_2(\mathbf{s}) \tag{7}$$

where $\zeta_2(\mathbf{s})$ comes from a spatially-structured process and $\varepsilon_2(\mathbf{s}) \sim N(0, \tau_2^2)$. Similarly, the log-precipitation model output is modeled using the latent process $\widetilde{X}_2(B)$

$$P_2(B) = \mu_2 + \widetilde{X}_2(B) + \eta_2(B) \tag{8}$$

where an overall mean $\mu_2$ occurs and each $\eta_2(B)$ is distributed as $N(0, \lambda_2^2)$. We specify a common latent process, $\widetilde{X}_2(B)$, that drives both the probability of precipitation occurrence and the rainfall accumulation, for both observed data (at grid cells containing $\mathbf{s}$) and model output (at all grid cells). This is a feature of the model that allows to smooth some unrealistic zero values (due to measurement error) in the middle of a storm [7].

## 2.1 Beyond the First Stage of Modeling

Dependence between temperature and precipitation might be introduced through the spatial latent processes, both at point and areal resolution. In the first case, we can anticipate association between the spatial latent processes $\zeta_1(\mathbf{s})$ and $\zeta_2(\mathbf{s})$ via the coregionalization method, that is assuming a linear combination of two independent univariate zero-mean unit variance spatial Gaussian processes. The association between the latent spatial processes induces a correlation to the response variables $Y_1(\mathbf{s})$ and $Y_2(\mathbf{s})$ [6]. The independence between the latent processes $\zeta$'s can be achieved using a diagonal coregionalization matrix.

In the second case, we focus on the latent processes $\widetilde{X}_1(B)$ and $\widetilde{X}_2(B)$. Since temperature and precipitation output arise jointly from the numerical model they are not independent. To accommodate correlation and promote spatial smoothing, we specify a bivariate intrinsic conditional autoregressive (BICAR) process for $\widetilde{X}_1(B)$ and $\widetilde{X}_2(B)$ in order to introduce multiple, dependent spatial latent effects associated with grid cells. The BICAR process is appealing because it links the temperature and precipitation output through a correlation parameter $\rho$ which has easy interpretation. Through (2) and (8) and the BICAR specification for the grid-level latent processes, we both account for the uncertainty in the numerical model and handle the spatial misalignment between the point-level observations and the grid-referenced model output. Indeed, we are implicitly linking the observed weather variables to the model output at all the neighboring cells of the grid cell containing the site.

The hierarchy of our model is completed by specifying the prior distributions of all hyperparameters. Noninformative conjugate priors are assumed. In particular, we place inverse gamma priors $IG(2, 1)$ for all variance parameters, implying that these variance components have prior mean 1 and infinite variance. Normal distributions $N(0, 10^3)$ are assumed as priors for all regression coefficients. The model is fitted using the Markov Chain Monte Carlo (MCMC) algorithm; a Gibbs sampling scheme is used to draw samples from the joint posterior distribution. Spatial predictions are based upon the posterior predictive distribution and sampled by composition.

## 3 Application to Emilia-Romagna Data

We illustrate our approach by combining the observed temperature and precipitation with the output from a numerical model for such variables over the Emilia-Romagna Region (Italy). Figure 1 shows the daily mean temperature in °C (left panel) and daily accumulated precipitation in mm (right panel) collected by 157 weather stations on March 5th, 2014. Even if it was a rainy day, more than 15% of sites show zero precipitation values. In general, there might be separate locations for each variable, but here we restrict ourselves to co-located data. We set aside 35 sites for model validation.

As a second data source, the numerical weather prediction model COSMO-I7 [2] is available. COSMO-I7 is the Italian version of the non-hydrostatic limited-area atmospheric prediction model developed by the European Consortium for Small-scale Modeling (www.cosmo-model.org) and operated by the Hydrometeorological Service of the Regional Agency for Prevention, Environment and Energy (ARPAE) of Emilia Romagna. It produces forecasts at 7 km resolution for several meteorological variables. Figure 2 shows the daily mean temperature in °C (left panel) and daily accumulated precipitation in mm (right panel) estimated by COSMO-I7 on March 5th, 2014 over the region. The precipitation value for more than 20% grid cells is zero.

Temperature and precipitation come jointly from COSMO-I7 model where several deterministic equations describe the relationship between the two variables. As a result, the output of the two variables can not be considered independent. Since for observed data descriptive analysis shows no residual correlation between temperature and precipitation given the corresponding COSMO-I7 output, it is reasonable to assume that temperature and precipitation are conditionally independent given the COSMO-I7 output. In other words we assume $\zeta_1(\mathbf{s})$ and $\zeta_2(\mathbf{s})$ independent spatial processes, each equipped with an exponential correlation function, i.e. $Cov(\zeta_k(\mathbf{s}), \zeta_k(\mathbf{s}')) = \exp(-\phi_k||\mathbf{s} - \mathbf{s}'||)$, where for $k = 1, 2$, $\phi_k$ is the spatial decay parameter fixed at roughly the 80 and 40% of the maximum distance between sites for temperature and precipitation, respectively.

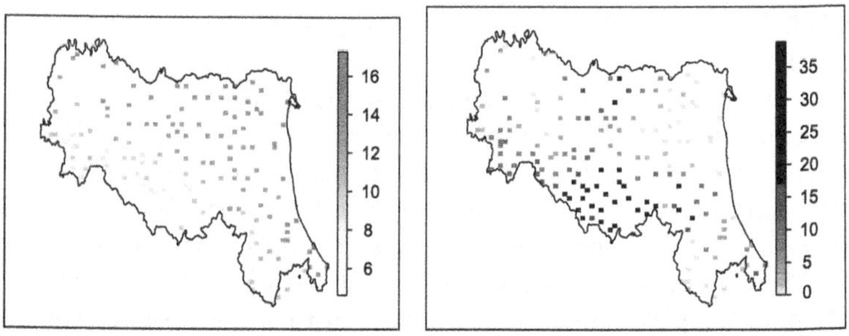


**Fig. 1** Daily mean temperature in °C (left panel) and daily accumulated precipitation in mm (right panel) observed on March 5th, 2014 over Emilia-Romagna region

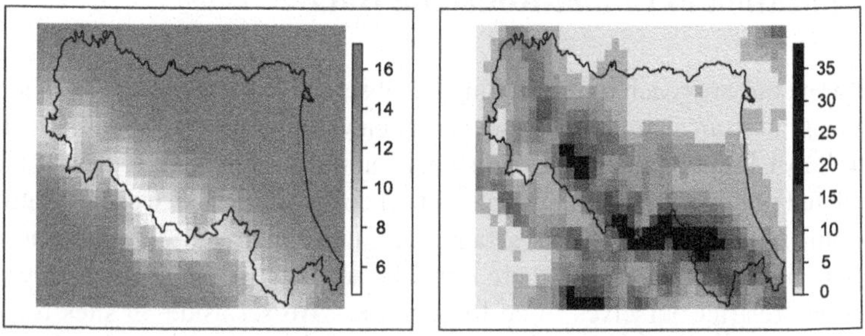


**Fig. 2** Daily mean temperature in °C (left panel) and daily accumulated precipitation in mm (right panel) estimated by COSMO-I7 on March 5th, 2014 over Emilia-Romagna region

## *3.1 Results*

Posterior summaries of all model parameters are shown in Table 1. Intercept parameters $\beta_0$ and $\mu_1$ of temperature models have similar posterior estimates with higher credible intervals associated with observed data since there are much less sites than model output grid cells. Similar comments apply to posterior summaries of $\gamma_0$ and $\mu_2$ and $\alpha_0$ and $\mu_2^*$. Posterior summaries of slope parameter $\beta_1$ show good agreement between monitoring data and COSMO-I7 output for temperature, while the slope parameter for precipitation, $\gamma_1$, is not significant. The parameter $\alpha_1$ is positive acknowledging that higher rain accumulation corresponds to higher probability of precipitation, as expected. Finally, the parameter $\rho$ shows a significant negative correlation between temperature and precipitation COSMO-I7 output.

Figures 3 and 4 show the posterior predictive maps of temperature and precipitation, each with the associated uncertainty map. Finally, Fig. 5 provides validation plots of observed data against both the COSMO-I7 output and out-of-sample predictions obtained from the model described in Sect. 2. Our joint predictions are closer to the observations relative to the raw COSMO-I7 output, showing how our joint stochastic model is able to calibrate the numerical model output.

**Table 1** Posterior summaries of model parameters

| | $\beta_0$ | $\beta_1$ | $\alpha_0$ | $\alpha_1$ | $\gamma_0$ | $\gamma_1$ | $\mu_1$ | $\mu_2$ | $\mu_2^*$ | $\sigma_1^2$ | $\sigma_2^2$ | $\lambda_1^2$ | $\lambda_2^2$ | $\tau_1^2$ | $\tau_2^2$ | $\rho$ |
|---|---|---|---|---|---|---|---|---|---|---|---|---|---|---|---|---|
| 2.5% | 11.76 | 0.76 | 0.76 | 0.14 | 0.59 | −0.54 | 13.74 | 0.95 | 0.80 | 0.30 | 0.99 | 0.02 | 0.02 | 0.53 | 0.13 | −0.29 |
| 50% | 12.73 | 0.88 | 1.01 | 0.37 | 1.52 | −0.27 | 13.75 | 0.99 | 0.91 | 0.67 | 1.52 | 0.03 | 0.03 | 0.71 | 0.23 | −0.23 |
| 97.5% | 13.75 | 0.99 | 1.26 | 0.61 | 2.44 | 0.01 | 13.76 | 1.04 | 1.02 | 1.53 | 2.19 | 0.03 | 0.03 | 0.93 | 0.38 | −0.16 |

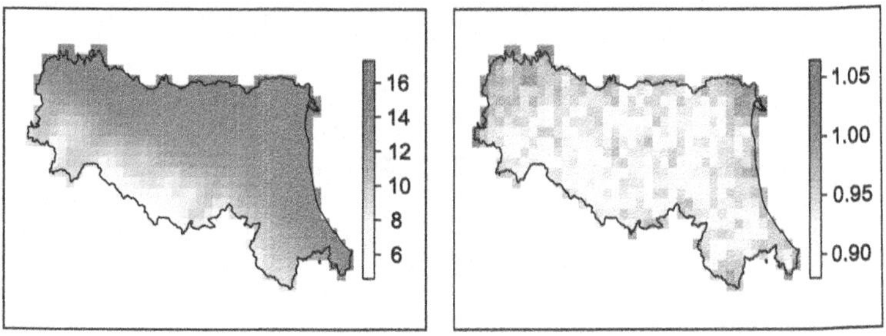


**Fig. 3** Posterior predictive map of temperature in °C (left panel) and corresponding standard deviation map (right panel) on March 5th, 2014 over Emilia-Romagna region

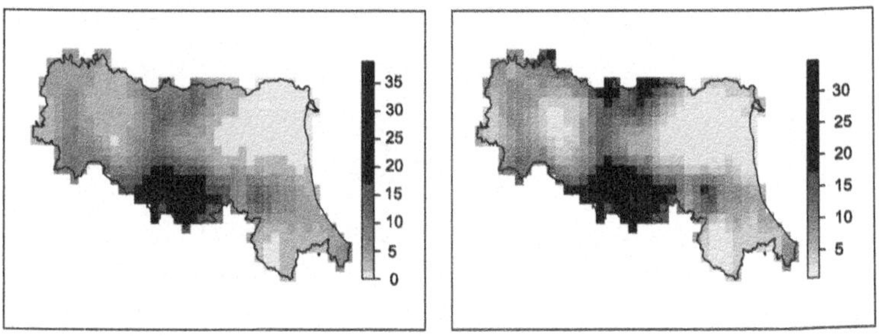


**Fig. 4** Posterior predictive map of daily precipitation in mm (left panel) and corresponding standard deviation map (right panel) on March 5th, 2014 over Emilia-Romagna region

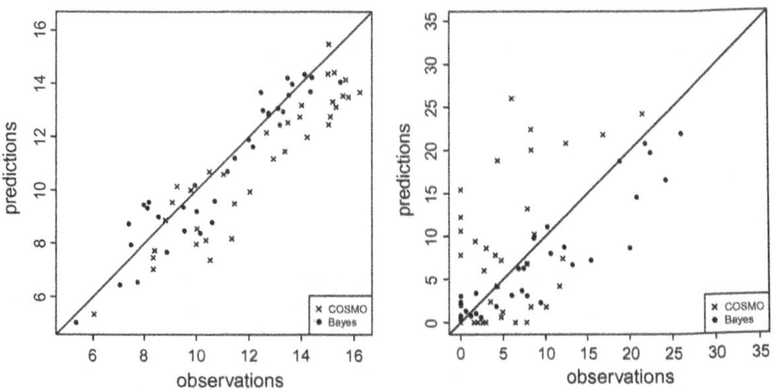


**Fig. 5** Temperature (left panel) and precipitation (right panel) validation plots

Future work will extend the model in order to accommodate for spatial data collected over time, i.e. space-time formulation of the joint downscaling model.

**Acknowledgements** Research was funded by MIUR through FIRB 2012 (project no. RBFR12URQJ) and PRIN 2015 (project no. 20154X8K23) grants. We also thank ARPAE-SIMC Emilia Romagna, for providing monitoring data set and numerical models' output.

## References

1. Albert, J.H., Chib, S.: Bayesian analysis of binary and polychotomous response data. J. Am. Stat. Assoc. **88**, 669–679 (1993)
2. Baldauf, M., Seifert, A., Förstner, J., Majewski, D., Raschendorfer, M., Reinhardt, T.: Operational convective-scale numerical weather prediction with the COSMO model: description and sensitivities. Mon. Weather Rev. **139**, 3887–3905 (2011)
3. Berrocal, V.J., Gelfand, A.E., Holland, D.M.: A spatio-temporal downscaler for output from numerical models. J. Agr. Biol. Environ. Stat. **14**, 176–197 (2010)
4. Berrocal, V.J., Gelfand, A.E., Holland, D.M.: Space-time data fusion under error in computer model output: an application to modeling air quality. Biometrics **68**, 837–848 (2012)
5. Bruno, F., Cocchi, D., Greco, F., Scardovi, E.: Spatial reconstruction of rainfall fields from rain gauge and radar data. Stoch. Environ. Res. Risk Asess. **28**, 1235–1245 (2013)
6. Chagneau, P., Mortier, F., Picard, N., Bacro, J.N.: A hierarchical bayesian model for spatial prediction of multivariate non-gaussian random fields. Biometrics **67**, 97–105 (2011)
7. Fuentes, M., Reich, B., Lee, G.: Spatial-temporal mesoscale modeling of rainfall intensity using gage and radar data. Ann. Appl. Stat. **2**, 1148–1169 (2008)
8. Neelon, B., Zhu, L., Neelon, S.E.: Bayesian two-part spatial models for semicontinuous data with application to emergency department expenditures. Biostatistics **16**, 465–479 (2015)

# Exploring Italian Students' Performances in the SNV Test: A Quantile Regression Perspective

**Antonella Costanzo and Domenico Vistocco**

**Abstract** Over the past decades, in educational studies, there is a growing interest in exploring heterogeneous effects of educational predictors affecting students' performances. For instance, the impact of gender gap, regional disparities and socio-economic background could be different for different levels of students' abilities, e.g. between low-performing and high-performing students. In this framework, quantile regression is a useful complement to standard analysis, as it offers a different perspective to investigate educational data particularly interesting for researchers and policymakers. Through an analysis of data collected in the Italian annual survey on educational achievement carried out by INVALSI, this chapter illustrates the added value of quantile regression to identify peculiar patterns of the relationship between predictors affecting performances at different level of students' attainment.

**Keywords** Quantile regression · INVALSI · Economic social and cultural status index

## 1 Introduction

The impact of educational factors on students' achievement is a theme of high interest in educational studies [6]. The classical approach focusing on average effects has been recently enhanced by the use of quantile regression (QR); among the others, see [5, 15]. QR permits to analyze the relationships between students characteristics, contextual variables, and educational achievements extending the viewpoint on the whole conditional distribution of performances. The technique appears particularly

A. Costanzo (✉)
National Institute for the Evaluation of Education System – INVALSI,
Via I. Nievo, 35, Roma, Italy
e-mail: antonella.costanzo@invalsi.it

D. Vistocco
Dipartimento di Economia e Giurisprudenza, Università degli Studi di Cassino e del Lazio Meridionale, Via S. Angelo S.N. – Località Folcara, Cassino, Frosinone, Italy
e-mail: vistocco@unicas.it



F. Mola et al. (eds.), *Classification, (Big) Data Analysis and Statistical Learning*, Studies in Classification, Data Analysis, and Knowledge Organization,
https://doi.org/10.1007/978-3-319-55708-3_13

interesting as educational achievements are often characterized by high degree of heterogeneity: for instance, the impact of family background, as well as of gender gap and regional disparities on learning outcomes could vary for the poorest performing pupils rather than for the best-performing ones. In such a framework, this chapter uses a QR approach on data collected by INVALSI within the national large-scale assessment test focusing on students' reading skills at the end of their eighth year of schooling in 2011–12 [7]. The aim is to investigate how individual characteristics and contextual factors might drive performances differently according to the level of ability. All computation and graphics were done in the R language [14] using the basic packages and the additional quantreg [8], and ggplot2 [16] packages.

The chapter is organized as follows: Sect. 2 describes the analyzed dataset, highlighting the main features of educational predictors and their relationship with the observed students' learning outcome; Sect. 3 briefly introduces quantile regression, and Sect. 4 presents the main results focusing on the estimated effects of predictors on students' performance with respect to the different level of students' abilities. Finally, concluding remarks and future avenues close in Sect. 5.

## 2 Students' Performances Data: Description and Main Evidence

The INVALSI National Survey on Educational Achievement (SNV) is on a census basis. It evaluates the whole population of students and schools at different stages of education (about 2,850,000 pupils in total and 15,000 schools). To reduce the probability of opportunistic behaviour from students and teacher in the standardized assessment test, in a subsample of schools (called *sample schools*) students take the test under the supervision of an external inspector [7]. In this chapter, the analysis is carried out on a random sample consisting of 23,035 Italian students at their eighth year of education (out of 587,412 students belonging to *sample schools*) in 2011–12. The normalized test score in reading comprehension test is used as a measure of students' ability defined in the range [0, 100]. Individual information concerning students' characteristics are gender, the family background expressed by the Economic Social and Cultural Status Index (ESCS) [11], regional areas (north, centre and south Italy), citizenship (native, non-native), and school attendance (regular, advanced and late student).

Table 1 reports the main summary statistics concerning the pupils' test score: it emerges as a skewed distribution, confirmed by the density plot of Fig. 1, where

**Table 1** Main summaries for the students' test score in reading comprehension

| | min | $q_1$ | *median* | $q_3$ | mean | sd | max | skew | kurt |
|---|---|---|---|---|---|---|---|---|---|
| Test score | 5.88 | 64.71 | 75.29 | 85.53 | 73.30 | 13.95 | 100 | −0.79 | 0.43 |

the score test is represented for the three geographical areas. Table 2 reports the summaries (columns) for the test score computed for the groups associated with the levels (rows) of the selected covariates. As expected, regular students obtain better results than late students, on average. Also, females show higher reading achievements score than males while non-native students lag behind native students. Geographical differences are also evident: students from the south obtain, on average, lower results than students from the north and the centre. The differences in the distribution of students' performance with respect to regional areas are also evident from the analysis of the three densities in Fig. 1. This result was quite expected; in fact, numerous studies have recorded a significant gap between southern and northern regions in Italy [7, 12]. However, by analyzing Fig. 1 and Table 2, it seems that regional differences can be better appreciated in terms of distributional changes of performances rather than in terms of average changes. In particular, while differences in average scores seem not particularly pronounced, the distribution of performances pertaining to students from the south appears slightly more left skewed than the remaining others. In this setting, focusing exclusively on changes in the means might misestimate the impact of geographical factor on students' performance.

Figure 2 represents the boxplots of the students test score distribution in the three regional areas according to the different level of students' socio-economic background. As expected, students with good living standard conditions are more likely to obtain higher results [3]. However, such a relationship tends to decrease as long as students' socio-economic background increases. This is particularly evident in the northern and in the central regions. Interestingly, regional disparities in the observed performance also matter among students with poorer living conditions; in fact, in

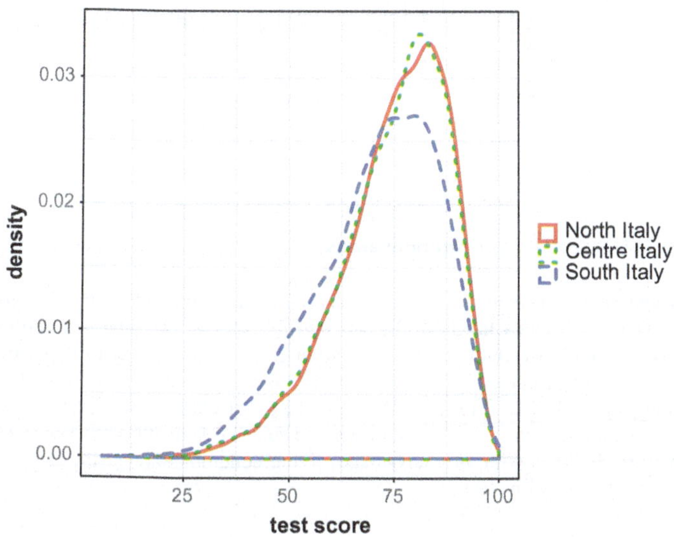


**Fig. 1** Distribution of students' performance in reading comprehension by regional areas

**Table 2** Main summaries for the students' performance in reading skills computed for the groups associated to the levels of the selected covariates

| | mean | sd | min | $q_1$ | *median* | $q_3$ | max |
|---|---|---|---|---|---|---|---|
| Male | 72.82 | 13.87 | 11.76 | 64.71 | 75.29 | 83.53 | 100.00 |
| Female | 73.79 | 14.01 | 5.88 | 65.88 | 76.47 | 84.70 | 100.00 |
| Native | 74.17 | 13.43 | 9.41 | 65.88 | 76.47 | 84.70 | 100.00 |
| Non-native | 65.34 | 15.98 | 5.88 | 54.11 | 67.05 | 77.64 | 97.64 |
| Regular student | 74.62 | 13.20 | 9.41 | 67.05 | 76.47 | 84.70 | 100.00 |
| Advanced student | 74.72 | 14.47 | 17.65 | 67.06 | 77.65 | 84.71 | 98.82 |
| Late student | 63.30 | 15.22 | 5.88 | 54.11 | 64.70 | 74.11 | 97.64 |
| North Italy | 75.00 | 13.03 | 18.82 | 67.06 | 76.47 | 84.71 | 100.00 |
| Centre Italy | 74.53 | 13.29 | 11.76 | 67.06 | 77.65 | 84.71 | 100.00 |
| South Italy | 70.88 | 14.84 | 5.88 | 61.17 | 72.94 | 82.35 | 100.00 |

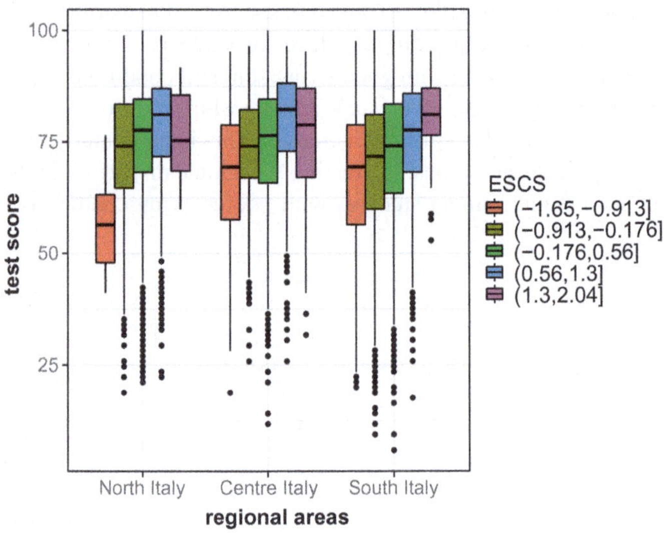


**Fig. 2** Distribution of test score by regional areas according to different levels of socio-economic background. The ESCS index has been divided into five classes of equal size representing the different students' living standard conditions. The first class corresponds to very disadvantaged students with ESCS value in the range (–1.65, –0.913]; the second class pertains to disadvantaged students in the range (–0.913, –0.176], the third and the fourth classes correspond to students with average socio-economic status (–1.176, 0.56] and advantaged students with ESCS Index in the range (0.56, 1.3]. Students with very advantages living conditions are in the range (1.3, 2.04], which corresponds to the fifth class

the northern regions students achieve the lowest performance whereas students from the centre and from the south are able to overcome their disadvantaged starting point: they indeed obtain higher test scores compared to students from the north. This phenomenon is known in the literature as *resilience* and it represents the ability of disadvantaged students to obtain high achievement scores [1]. Data suggest that resilient pupils mostly come from the south, are Italian and there is no prevalence of males or females. Further researches on this topic can be found in [13].

Undoubtedly, assessing how the combination of individual variables and contextual factors characterizes students' performance and whether it varies according to the level of pupils' ability could be a good instrument to report the findings of a national assessment even from a descriptive point of view. Starting from this consideration, in future applications, it might be interesting to estimate, through a QR approach, the combined effect of geographic factor or gender gap and socio-economic status along the entire conditional distribution of students' achievement. This idea goes behind the scope of this study which is to provide a fine-grained picture of the differential effects of the selected covariates on learning outcomes considering as first task-only additive effects.

## 3 Quantile Regression: The Essentials

Quantile regression [10] may be viewed as an extension of least squares estimation of conditional mean models to the estimation of an ensemble of models for several conditional quantile functions, taking into account the effects a set of covariates plays on a response variable. Since multiple quantiles can be modeled, it is possible to achieve a more complete understanding of how the response distribution is affected by predictors catching information about changes in location, spread, and shape [4, 9].

In analogy with the classical linear regression framework, a linear regression model for the $\theta$–th conditional quantile of $y_i$ can be expressed as

$$\mathbf{Q}_{y_i(\theta)|\mathbf{x}_i} = \mathbf{x}_i^T \boldsymbol{\beta}_\theta,$$

where $y$ is a scalar-dependent variable, $\mathbf{x}_i^T$ is the $k \times 1$ vector of explanatory variables, $\boldsymbol{\beta}$ is the coefficient vector, $\theta$ is the conditional quantile of interest. With respect to classical linear regression methods, based on minimizing sums of squares residuals, quantile regression methods are based on minimizing asymmetrically weighted absolute residuals:

$$\min_{\boldsymbol{\beta}} \sum_{y_i \geq \mathbf{x}_i^T \boldsymbol{\beta}} \theta|y_i - \mathbf{x}_i\boldsymbol{\beta}| + \sum_{y_i < \mathbf{x}_i^T \boldsymbol{\beta}} (1-\theta)|y_i - \mathbf{x}_i\boldsymbol{\beta}|$$

By setting $\theta = 0.5$, the previous equation provides the median solution, while the use of any $\theta$ between 0 and 1 allows to study the dependence structure at any location

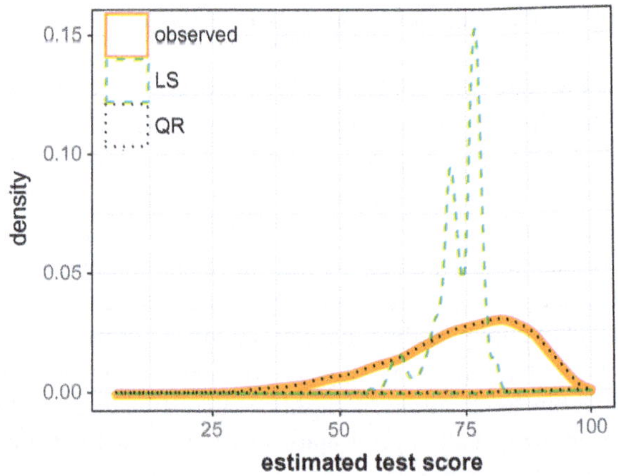


**Fig. 3** Distribution of observed students' performances (straight line) and of the estimated performances distribution using LS (dashed line) and the QR approach (dotted line)

of the response conditional distribution. The estimated $\hat{\boldsymbol{\beta}}_\theta$ in QR linear models have the same interpretation as those of any other linear model: each $\hat{\beta}_\theta$ coefficient can be interpreted as the rate of change in the $\theta$-th quantile of the dependent variable distribution per one unit change in the value of the corresponding regressor, holding constant the others. The study of the effect played by the considered regressors on the whole conditional distribution and not only on the conditional mean is useful to obtain a complete estimation of the response variable. To this end, Fig. 3 enlightens the added value of QR. The QR-estimated density (dotted line) for data concerning students' performances is, indeed, pretty much equivalent to the observed density (solid line), while it is evident the approximation obtained focusing only on the conditional mean (dashed line). For details about the main approaches for density estimation in QR, see [4, 9]. Next section introduces the QR model for analyzing the students' performances mainly focusing on the interpretation of the QR coefficients.

## 4 The Effects of Educational Predictors on Italian Students' Performances: QR Results

The Italian students' performance is regressed on gender, ESCS indicator, regional area, citizenship and school attendance, using the ordinary least squares regression (LS) and the QR approach, for $\theta = (0.1, 0.25, 0.5, 0.75, 0.9)$. The estimated coefficients for each regressor are shown in Table 3 and depicted in Fig. 5. Before interpreting the results, it is worth to notice that the LS and the QR models are estimated using a quite large dataset. Although some caution in interpreting the size and the significance of the results is required, this chapter might contribute to assess the potential relevance and the methodological adequacy of a QR approach in dealing with large-scale data set which constitutes one of the recent concerns among researchers in education [2].

**Table 3** LS and QR estimates for reading comprehension. Covariates: gender (ref: male), school attendance (ref: regular), citizenship (ref: native), regional area (ref: north Italy), ESCS. Standard errors were computed using the bootstrap (500 reolications). In bold the estimates statistically significant (p-values < 0.05)

| | LS | QR | | | | |
|---|---|---|---|---|---|---|
| | | $\theta = 0.1$ | $\theta = 0.25$ | $\theta = 0.5$ | $\theta = 0.75$ | $\theta = 0.9$ |
| (Intercept) | 76.34 | 60.46 | 69.44 | 77.92 | 85.05 | 89.87 |
| Female | **0.54** | **0.13** | **0.66** | **0.79** | **0.71** | **0.48** |
| ESCS | **3.65** | **4.87** | **4.51** | 4.00 | **2.86** | **1.95** |
| Immigrant | **–6.51** | **–10.99** | **–8.49** | **–6.84** | **–4.39** | **–2.69** |
| Advanced student | 0.89 | –1.10 | 0.56 | 1.12 | 1.21 | 1.48 |
| Late student | **–9.24** | **–10.70** | **–11.13** | **–9.99** | **–8.52** | **–6.39** |
| Centre–Italy | **–0.76** | **–1.33** | **–0.98** | **–0.65** | **–0.59** | **–0.47** |
| South–Italy | **–4.14** | **–7.97** | **–5.53** | **–3.82** | **–2.41** | **–1.17** |

The QR model has been estimated using five conditional quantiles, representing the different levels of students' abilities, i.e. $\theta = (0.1, 0.25, 0.5, 0.75, 0.9)$. Although it would be possible to obtain estimates across the entire interval of conditional quantiles, this chapter focuses only on five quantiles representing five different levels of students' abilities. Such quantiles can be interpreted consistently with the levels of reading achievement measurement scale largely adopted in the educational literature [3, 7, 12]. Moreover, the use of a limited number of quantiles represents a rightful compromise between the amount of output to manage and the results to interpret. The graphical visualization allows to appreciate the pattern of QR estimates in comparison with the LS coefficients. For instance, QR results highlight that the gender gap between males and females in reading achievement tends to be wider among lower performers (around the 25th percentile) rather than for higher performers. Citizenship is an additional factor affecting students' performance: on average, immigrants obtain lower performance than natives; this result is also in line with findings from larger scale surveys using national students' data [7, 12]. Notwithstanding, QR estimates clearly illustrate that the negative impact of being immigrant becomes less influential moving from lower to upper quantiles, namely from lower to higher achievers. Next, school attendance influences students' final outcome. Intuitively, being a late student exerts a negative effect on learning outcomes compared to regular students. However, such a negative impact becomes narrower moving from lower level to higher level of students' abilities. QR results highlight also the impact of the geographical factor on the test score distribution: students' from the south obtain lower performance than students from the north for each level of ability. However, the intensity of regional disparities is higher for low achievers rather than for high achievers. As regards the relationship between individual living conditions (ESCS) and students' achievement, QR estimates depict a positive but differential effect of ESCS between low performers and high performers. In particular, the major impact can be appreciated among students with lower level of skills which often come from

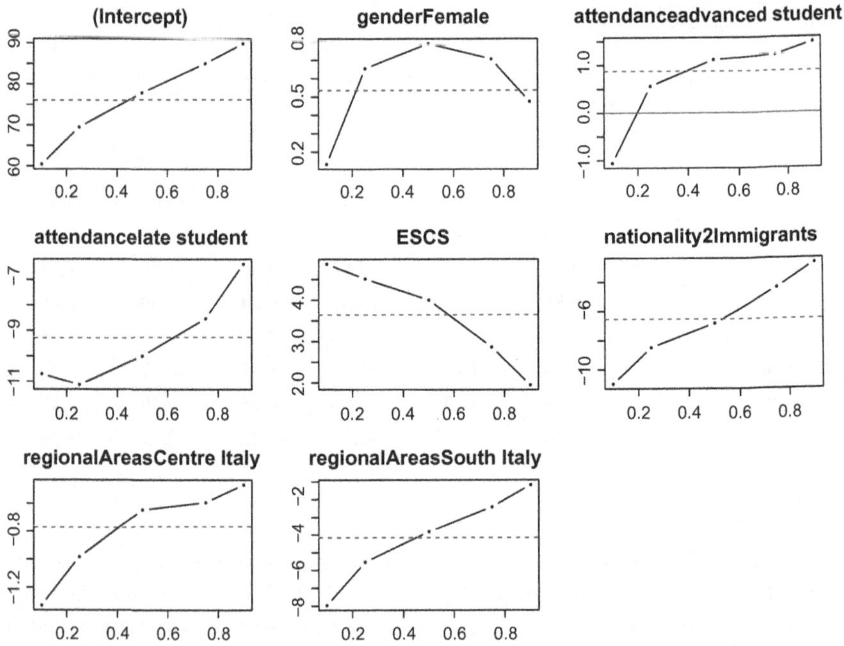


**Fig. 4** LS and QR regression coefficients for the different predictors of students' ability. The horizontal axes display the quantiles while the estimated effects are reported on the vertical axes. The horizontal straight line parallel to the x-axis corresponds to LS coefficients, while the dashed line represents QR estimates

poorer economical conditions, hence with a limited access to educational resources (Fig. 4).

As already pointed out, descriptive results suggest that the relationship between socio-economic background and students' performance varies according to students' skills and with respect to geographical areas (see Fig. 2). Hence, assessing the combined effect "beyond the mean" on the test score distribution might be interesting for future studies using the INVALSI data. As a preliminary basis for discussion, Fig. 5 represents the estimated ESCS effect on students' learning outcome at five quantiles (the five lines) corresponding to the different levels of students' ability, according to geographical areas (the three different symbols). It turns out that, in southern regions, poorer living standard conditions seem to exert a major negative impact on low achievers while it becomes less intensive moving from lower to upper quantiles of the conditional test score distribution. On the other hand, moving from poorer to better living standard conditions it emerges that the potential impact of higher values of ESCS on educational attainment is less effective as the level of students' ability increases. This pattern is particularly evident in northern and central areas characterized by relatively higher values of ESCS and with a small evidence of resilience compared to the southern regions counterpart.

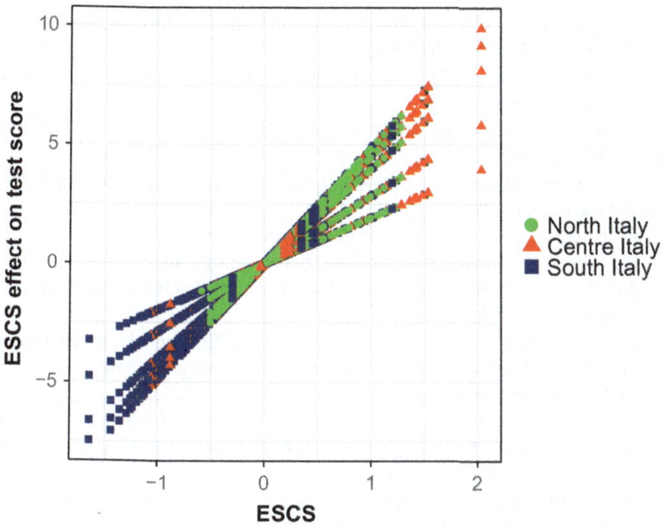


**Fig. 5** ESCS effects on test score at different level of students' ability by geographical areas

## 5 Concluding Remarks and Future Avenues

The chapter illustrates through an analysis of real data, the added value of quantile regression to obtain information about the role of educational predictors in driving students' performances at different levels of ability. The main findings suggest that, compared to the average effect, the gap in favour of females in reading skills is significantly wider among low achievers. Also, the impact of regional disparities and family background can be better appreciated at the extremes of the conditional distribution of performances. Since the evaluation of student learning outcomes is considered increasingly necessary for monitoring and improving education quality, QR information might be used from teachers and policymakers to move towards a more effective educational policy implementation.

## References


1. Agasisti, T.: Does competition affect schools' performance? Evidence from Italy through OECD-PISA data. Eur. J. Educ. **46**(4), 549–565 (2011)
2. Chen, F., Chalhoub-Deville, M.: Principles of quantile regression and an application. Lang. Test. **31**(1), 549–565 (2011)
3. Coleman, J.S., Campbell, E.Q., Hobson, C.J., McPartland, J., Mood, A.M., Weinfall, F.D., York, R.L.: Equality of Educational Opportunity. Office of Education, US National Center for Education Statistics, OE Series (1966)
4. Davino, C., Furno, M., Vistocco, D.: Quantile Regression: Theory and Applications. Wiley Series in Probability and Statistics (2013)

5. Eide, E., Showalter, J.: The effect of school quality on student performance: a quantile regression approach. Econ. Lett. **5**, 345–350 (1998)
6. Hanushek, E., Woessmann, L.: The role of cognitive skills in economic development. J. Econ. Lit. **46**(3), 607–668 (2008)
7. INVALSI: Rilevazioni nazionali sugli apprendimenti 2011–2012, INVALSI. http://www.invalsi.it/snv2012/ (2012)
8. Koenker, R.: quantreg: Quantile Regression. R package version 5.29. https://CRAN.R-project.org/package=quantreg (2016)
9. Koenker, R.: Regression quantiles. Econometrica **46**(1), 33–50 (1978)
10. Koenker, R., Basset, G.: Regression quantiles. Econometrica **46**(1), 33–50 (1978)
11. OECD: Education at a Glance 2007: OECD Indicators. PISA, OECD Publishing, Paris. http://www.oecd.org/education/skills-beyond-school/39313286.pdf (2007)
12. OECD: Low-Performing Students: Why They Fall Behind and How To Help them Succeed. PISA, OECD Publishing, Paris. http://www.oecd.org/edu/low-performing-students-9789264250246-en.htm (2012)
13. OECD: Against the Odds: Disadvantaged Students Who Succeed in School. OECD Publishing, Paris. www.oecd.org/edu/school/programmeforinternationalstudentassessmentpisa/ (2011)
14. R Core Team: R: A Language and Environment for Statistical Computing. R Foundation for Statistical Computing, Vienna, Austria. https://www.R-project.org/ (2016)
15. Tzavidis, N., Brown, J.: Using M-quantile models as an alternative to random effects to model contextual value added of schools in London. Leading Education and Social Research, Institute of Education, University of London. http://repec.ioe.ac.uk/REPEc/pdf/qsswp1011.pdf (2010)
16. Wickham, H.: ggplot2: Elegant Graphics for Data Analysis. Springer, New York (2009)

# Estimating the Effect of Prenatal Care on Birth Outcomes

**Emiliano Sironi, Massimo Cannas and Francesco Mola**

**Abstract** Using data from official hospital abstracts on deliveries occurred in Sardinia during the years 2010 and 2011, we implemented an Augmented Inverse Probability Weighted (AIPW) model in order to study the effect of increased prenatal care during pregnancy on birth outcomes. Results showed that moderate levels of prenatal care, as measured by the number of sonograms, increase the Apgar score of the infant, while a higher number of sonograms does not have any additional marginal effect on the outcome.

**Keywords** Childbirth outcomes · Treatment Effect · IPW and AIPW models

## 1 Introduction

During the 1990s support grew for the hypothesis that prenatal care is linked to better birth outcomes. In 1988 the U.S. Office of Technology Assessment stated [9]:

> The weight of the evidence on the effectiveness of prenatal care supports the contention that birth outcomes can be improved with earlier or more comprehensive prenatal care. The evidence appears to support the value of both early and frequent prenatal care and the provision of enhanced services to adolescents and high-risk women.

However, today, there is no clear consensus on the relationship between prenatal care and improvements in birth outcomes [1]. Do variations in such care play a

E. Sironi (✉)
Dipartimento di Scienze Statistiche, Università Cattolica del Sacro Cuore,
Largo Gemelli 1, Milan, Italy
e-mail: emiliano.sironi@unicatt.it

M. Cannas · F. Mola
Dipartimento di Scienze Economiche ed Aziendali, Università di Cagliari,
Viale Sant'Ignazio 83, Cagliari, Italy

M. Cannas
e-mail: massimo.cannas@unica.it

F. Mola
e-mail: mola@unica.it

 
F. Mola et al. (eds.), *Classification, (Big) Data Analysis and Statistical Learning*,
Studies in Classification, Data Analysis, and Knowledge Organization,
https://doi.org/10.1007/978-3-319-55708-3_14

major role in the maternal and birth outcomes, or are these variations due to characteristics and behaviors which mothers and infants bring with them to the healthcare setting? Studies on this subject cannot use data from randomized controlled trials and this factor raises the possibility that results are biased with the risk to over- or underestimate the true effect of prenatal care on newborn health outcomes. Few non-randomized (observational) studies control for the self-selection bias of women [6, 8]. Women most likely to have the best birth outcomes are also those who have the highest number of prenatal care visits. In response, healthcare providers may require more prenatal visits than is typically recommended for low-risk pregnancies. In this situation, any cross-sectional relationship between birth outcomes and visits would understate the benefits of care [11].

The aim of this chapter is to measure whether high levels of prenatal care, as measured by the number of sonograms (scans) undergone by the mother during pregnancy, influence the health of the child. Literature provides details on suitable measures used as proxy of childbirth outcomes: Piper et al. [10] and Racine et al. [11] mainly focused on the weight of children, while Salustiano et al. [12] and Villar et al. [14] used Apgar score measured five minutes after delivery. Although controversial in literature, 5-min Apgar scores of less than 7 were associated with cognitive impairment as measured by academic achievement at 16 years of age [12]. The Apgar scale is determined by evaluating the newborn baby on five criteria on a scale from zero to two, then summing up the five values thus obtained. The resulting Apgar score ranges from 0 to 10. The five criteria are summarized using words chosen to form a backronym: Appearance, Pulse, Grimace, Activity, and Respiration. Following Thacker [13], we used the number of sonograms for summarizing prenatal care. In literature, we may observe also alternative measures for medicalization, taking into account the number of visits [12]. However, because of visits and ultrasounds happen in the same circumstances [2], the choice of either variable is indifferent.

After this brief introduction the reminder of the paper is organized as follows: Sect. 2 presents data and provides summary statistics about the main variables used in the analyses. In Sect. 3 we illustrate the model implemented while Sect. 4 displays the results. Section 5 concludes.

## 2 Data

We considered a data set containing information on deliveries occurred in 20 hospitals of the Italian region of Sardinia during the 2-year period 2010–2011. The data set collects individual information contained in the "Certificato di Assistenza al Parto" (CeDAP), which is the official abstract designed for capturing clinical features and the social and demographic characteristics of the family.

Table 1 shows the frequencies of the dependent variable of the model, i.e., the Apgar scores. As shown in Table 1, about 95% of births are associated with Apgar

**Table 1** Frequencies of Apgar scores at 5 minutes after birth

| Apgar score (5 min) | Frequency | Percentage |
|---|---|---|
| $<7$ | 293 | 1.18 |
| 7 | 216 | 0.87 |
| 8 | 783 | 3.15 |
| 9 | 6,570 | 26.40 |
| 10 | 17,025 | 68.41 |
| Total | 24,887 | 100 |

**Table 2** Frequencies of sonograms during the whole pregnancy

| Number of sonograms | Frequency | Percentage |
|---|---|---|
| 0–3 | 4,364 | 17.54 |
| 4–5 | 4,964 | 19.95 |
| 6–7 | 5,829 | 23.42 |
| $\geq 8$ | 9,730 | 39.10 |
| Total | 24,887 | 100 |

scores equal to 9 or 10. This result displays high standard levels for birth outcomes in Sardinia. However, although birth outcomes are generally good, critical scores, associated to scores less than seven, are not absent in the descriptives.

Table 2 shows the distribution of the treatment variable, i.e., the number of sonograms. The model implemented belongs to the class of Inverse Probability Weighted (IPW) estimators used in literature for estimating the Average Treatment Effects (ATE). We used an augmented version (AIPW) of the classic IPW estimator, which was proposed by Cattaneo et al. [4] and it is suitable for estimating the effect of multi-valued treatment. More precisely, let $T = j$ be a multi-valued treatment with $j = 0 \ldots, J$. In our case, we assume that the treatment is the number of sonograms. The distribution of thc treatment levels is shown in Table 2.

As we can see from the Table 2, just under 40% of the women included in the sample experienced at least eight sonograms during the pregnancy, while an important minority (more than 17%) undergone low levels of medicalization (less than 4 sonograms during the whole pregnancy).

The reference category is set on 0–3 sonograms, because in Italy there is a policy for three routine ultrasound examinations, although 10% of the sample registers a level of medicalization below the prescribed threshold, as reported by Eurocat [5].

The high variability of the explanatory variable begs the question whether a change in the number of ultrasound examinations during pregnancy has any impact on birth outcomes. Hence, we are in presence of a treatment evaluation setting.

## 3 Model

We define the observed outcome of the variable of interest as given by

$$y_i = d_i(0)y_i(0) + d_i(1)y_i(1) + \cdots + d_i(J)y_i(J), \tag{1}$$

where $[y_i(0), y_i(1), \ldots, y_i(J)]'$ is an independent and identically distributed sample drawn from $[y(0), y(1), \ldots, y(J)]'$. The distribution of each $y(j)$ is the distribution of the outcome variable that would occur if individuals were given treatment level $j$. $di(j)$ is an indicator variable taking value 1 if unit i received treatment $j$ and the value 0 otherwise.

As our purpose is to evaluate the impact of the number of ultrasound on Apgar score, the quantity of interest in our setting is the ATE, defined as follows:

$$ATE = E[y(j)] - E[y(0)], \tag{2}$$

where $E[y(j)]$ is the expected birth outcome in case of treatment level $j = 1, ..., J$ and $E[y(0)]$ is the expected outcome in case of the baseline treatment level settled at $j = 0$. Since, we can observe only one of the $J$ treatment values for each unit $i$, ATE cannot be consistently estimated using sample means.

Indeed, individuals treated at different levels may differ from each other due to observable characteristics affecting both the probability of being treated at different levels and the outcome performances.

The main assumption of the model is based on classic identification assumptions, in particular "selection-on-observables" [7] requiring that

$$y(j) \perp d(j) | \mathbf{x} \tag{3}$$

That assumption implies that the distribution of each potential outcome $y(j)$ is independent on the random treatment $d(j)$, conditional on a vector of covariates $\mathbf{x}$, called "vector of confounders" including all the variables that may affect both the probability of being treated at level j and the corresponding outcome $y(j)$.

Augmented Inverse Probability Weighted (AIPW) estimators address the problem of selection on observables using weighted averages of the observed outcomes [3]. More in detail, the weights are the inverse of the estimated probability $p_j(\mathbf{x}_i)$ that an individual receives a treatment level j; $p_j(\mathbf{x}_i)$ is called generalized propensity score and has the characteristics of ranging in [0, 1]. This means that we are able to observe individuals of each covariate type in each treatment level. Obviously, the outcome of individuals who are likely to receive the treatment gets a weight close to one, while the outcome of individuals that are less likely to receive a treatment gets a weight much larger.

AIPW strategy models both the outcome and the treatment probabilities, obtaining double robust estimators [3]. Following Cattaneo [4], AIPW estimator can be obtained through the generalized method of moments. In particular, to describe AIPW estimator in the case of means, we introduce the following function:

$$e_j(\mathbf{x}_i; \mu_j) = E\{y_i(j) - \mu_j|\mathbf{x}_i\} = E\{y_i(j) - \mu_j|\mathbf{x}_i, T_i = j\}. \quad (4)$$

That is computed for each treatment level $j$. The AIPW estimator is constructed using the following moment condition for the mean:

$$E[\psi_{AIPW}\{\mathbf{z}_i, \mu_j, p_j(\mathbf{x}_i), e_j(\cdot, \mu_j)\}] = 0 \quad (5)$$

$$\psi_{AIPW}\{\mathbf{z}_i, \mu_j, p_j(\mathbf{x}_i), e_j(\cdot, \mu_j)\} = \frac{d_i(j)}{p_j(\mathbf{x}_i)} - \frac{e_j(\mathbf{x}_i, \mu_j)}{p_j(\mathbf{x}_i)}\{d(j) - p_j(\mathbf{x}_i)\}. \quad (6)$$

The first term of $\psi$ is the IPW estimator, while the term that follows the minus sign is an additional component that gives the estimator an optimal property called "double robustness": although we consistently estimate $p_j(\mathbf{x}_i)$ and $e_j(\mathbf{x}_i, \mu_j)$ with their sample counterparts, the AIPW estimators require only that either $p_j(\mathbf{x}_i)$ or $e_j(\mathbf{x}_i, \mu_j)$ be correctly specified. In this sense vector $\mathbf{x}_i$ in $p_j(\mathbf{x}_i)$ may be different from the specification of vector $\mathbf{x}_i$ in $e_j(\mathbf{x}_i, \mu_j)$.

The conditional expectation displayed above can be estimated from observational data, resolving the following equation as follows:

$$\hat{\mu}_{AIPW,j} \text{ such that } \frac{1}{n}\sum_{i=1}^{n} \psi_{AIPW}\{\mathbf{z}_i, \hat{\mu}_{AIPW,j}, \hat{p}_j(\mathbf{x}_i), \hat{e}_j(\cdot, \hat{\mu}_{AIPW,j})\} = 0 \quad (7)$$

Hence, we specified a set of covariates as predictors of the treatment such as the age of the mother, her nationality (Italian or not), her education (ordered into four categories: primary, secondary, or tertiary education and a residual category collecting missing values), a dummy variable for indicating nulliparous women and another one for distinguishing smoking behavior. Moreover, we added a dummy for taking into account an adverse negative progress of the pregnancy; the outcome equation includes infant's weight, mother's gestational age (categorized into preterm births, normal-term, and late-term), smoking behavior, education, nationality, and age.

## 4 Results

Estimates of ATE are displayed in Table 3: lowest levels of prenatal care are associated with worst performances in Apgar scores. Women that underwent more than three sonograms during pregnancy performed better in terms of Apgar scores than those belonging to the reference category. In particular, the best birth outcomes are associated with women doing between 6 and 7 sonograms, while not any significant increase in relative performances was found for those doing more than 7 scans. Although about 40% of women experienced at least eight ultrasound examinations during the whole pregnancies, empirical evidence shows that childbirth outcomes

**Table 3** Estimated ATE of the number of sonograms on the Apgar score

| Number of sonograms | ATE | Robust Std. Err. | $z^a$ values |
|---|---|---|---|
| 4–5 (vs. 0–3) | 0.109 | 0.023 | 4.739 *** |
| 6–7 (vs. 0–3) | 0.177 | 0.022 | 8.045 *** |
| ≥8 (vs. 0–3) | 0.167 | 0.020 | 8.350 *** |

[a] *** indicates significance at 0.001 level; ** at 0.01 level; * at 0.05 level

do not improve further after overcoming the threshold of 7. Results are robust also considering little changes in the categorization of the number of scans.

## 5 Conclusions

Results show that the relationship between the number of sonograms and Apgar score is not monotone: medium levels (between 4 and 7 scans) instead of lowest levels of prenatal care (between 0 and 3) produce high increases in the Apgar performances but a higher number of sonograms do not have any positive additional effect on the outcome. Results are useful also for developing further studies in order to obtain policy implications about best medical practice. A recommended practice has to take into account a correct balance between the need for prenatal care and prevent unnecessary excesses in medical assistance.

**Acknowledgements** The authors would like to thank *Regione Autonoma della Sardegna* for providing the anonymized data used in the analysis.

## References

1. Alexander, G.R., Kotelchuck, M.: Assessing the role and effectiveness of prenatal care: history, challenges, and directions for future research. Public Health Rep. **116**(4), 306–316 (2001)
2. Bennet, M.J.: Routine ultrasound and the gynaecology visit. Curr. Opin. Obstet. Gynecol. **10**(5), 387–390 (1998)
3. Cattaneo, M.D.: Efficient semiparametric estimation of multi-valued treatment effects under ignorability. J. Econometrics **155**, 138–154 (2010)
4. Cattaneo, M.D., Drukker, D.M., Holland, A.D.: Estimation of multivalued treatment effects under conditional independence. Stata J. **13**, 407–450 (2013)
5. Eurocat Special Report: Prenatal Screening Policies in Europe. http://www.orpha.net/actor/Orphanews/2010/doc/Special-Report-Prenatal-Screening-Policies.pdf (2010)
6. Joyce, T.: Impact of augmented prenatal care on birth outcomes of Medicaid recipients in New York City. J. Health Econ. **18**(1), 31–67 (1999)
7. Holland, P.W.: Statistics and causal inference. J. Am. Stat. Assoc. **81**(396), 945–960 (1986)
8. Liu, G.G.: Birth outcomes and the effectiveness of prenatal care. Health Serv. Res. J. **32**(6), 805–823 (1998)
9. Office of Technology Assessment, U.S. Congress.: Healthy children: investing in the future. U.S. Government Printing Office, Washington, D.C. (OTA-H-345) (1988)

10. Piper, J.M., Mitchel Jr., E.F., Ray, W.A.: Evaluation of a program for prenatal care case management. Fam. Plann. Perspect. **28**, 65–68 (1996)
11. Racine, A.D., Joyce, T.J., Grossman, M.: Effectiveness of health care services for pregnant women and infants. Fam. Plann. Perspect. Future Child. **2**(2), 40–57 (1992)
12. Salustiano, A.M.A., Bonini Campos, J.A.D., Ibidi, S.M., Ruano, R., Zugaib, M.: Low Apgar scores at 5 minutes in a low risk population: maternal and obstetrical factors and postnatal outcome. Revista da Associacao Médica Brasileira **58**(5), 587–593 (2012)
13. Thacker, S.B.: Quality of controlled clinical trials. The case of imaging ultrasound in obstetrics: a review. Br. J. Obstet. Gynaecol. **92**, 437–444 (1985)
14. Villar, J., Ba'aqeel, H., Piaggio, G., Lumbignon, P., Belizan, J.M., Farnot, U.: WHO Antenatal Care Trial Research Group. Lancet **357**(9268), 1551–1564 (2001)

# Part V
# Clustering and Classification

# Clustering Upper Level Units in Multilevel Models for Ordinal Data

**Leonardo Grilli, Agnese Panzera and Carla Rampichini**

**Abstract** We consider an explorative method for unsupervised clustering of upper level units in a two-level hierarchical setting. The idea lies in applying a density-based clustering algorithm to the predicted random effects obtained from a multilevel cumulative logit model. We illustrate the proposed approach throughout the analysis of data from European Social Survey about political trust in European countries.

**Keywords** Density-based clustering · Empirical Bayes predictions · European Social Survey · Proportional odds model · Random effects

## 1 Introduction

We consider the issue of clustering upper level units in a multilevel model with ordinal response. An example is the clustering of university courses (level 2 units) on the basis of the ratings expressed by students (level 1 units) using an ordinal scale, with the aim of identifying "good" and "bad" courses. Another example is the clustering of countries on the basis of individual satisfaction expressed on a Likert scale.

In two-level models, regardless of the nature of the response variable (continuous, binary, ...), the standard approach for clustering level 2 units is to assume a discrete distribution for the random effects and exploit non-parametric maximum likelihood [1] or multilevel latent class (or mixture) models [3, 15]. However, the selection of the number of mass points (latent classes), which corresponds to the number of clusters, is an open issue (traditional criteria are, for example, AIC and BIC,

L. Grilli (✉) · A. Panzera · C. Rampichini
Department of Statistics, Computer Science, Applications 'G. Parenti' University of Florence, Florence, Italy
e-mail: grilli@disia.unifi.it

A. Panzera
e-mail: a.panzera@disia.unifi.it

C. Rampichini
e-mail: rampichini@disia.unifi.it

 
F. Mola et al. (eds.), *Classification, (Big) Data Analysis and Statistical Learning*, Studies in Classification, Data Analysis, and Knowledge Organization,
https://doi.org/10.1007/978-3-319-55708-3_15

e.g., [8, 12]). Alternatively, the selection of the number of clusters can be incorporated in the estimation method, thus obtaining an unsupervised clustering procedure, like the EM-based approaches of [2] and [7], where the number of mass points is automatically selected depending on some *problem-driven* tuning parameters. However, these approaches are computationally demanding and the choice of tuning parameters is not trivial. Alternatively, the semi-parametric Bayesian approach based on the Dirichlet process [13] allows to jointly fit the model and select the number of clusters.

Given the difficulties in the implementation of the above methods, it is worth to enlarge the toolkit by developing simple exploratory unsupervised methods. To this end, we exploit the idea of density-based clustering, where the clusters are defined as high-density regions in the data space. In particular, for a two-level model with an ordinal response, we predict the random effects and then we exploit the resulting kernel density estimate, as an explorative tool, to identify clusters as high-density regions. Specifically, we use the density-based approach of Azzalini, Menardi, and Torelli [5, 9]. The novelty of our proposal is that density-based clustering is applied to predictions of unobserved quantities instead of observed variables. It is worth to note that, even if we focus on a multilevel model for an ordinal response, the proposed approach can be used for any kind of multilevel model.

## 2 Density-Based Clustering of Upper Level Units

A common choice for analyzing ordinal data with a hierarchical structure is the cumulative ordinal or proportional odds model [6]. This model characterizes the responses in $C$ ordinal categories in terms of $C-1$ cumulative logits. The covariate effects are assumed to be the same across the cumulative logits, and random effects are included in the model to capture the correlation in the responses of the level 1 units belonging to the same level 2 unit. The proportional odds model with random effects [6] is

$$\text{logit}(P(Y_{ij} \leq y_c | \mathbf{x}_{ij}, u_j)) = \alpha_c - \left(\boldsymbol{\beta}'\mathbf{x}_{ij} + u_j\right) \qquad c = 1, 2, \ldots, C-1, \quad (1)$$

where for $j = 1, 2, \ldots, J$ and $i = 1, 2, \ldots, n_j$, $Y_{ij}$ is an ordinal response, pertaining to the $i$th level 1 unit nested into the $j$th level 2 unit, with $C$ ordered categories $y_1, \ldots, y_C$; the $\alpha_c$s are strictly increasing category-specific intercepts, i.e., $\alpha_1 < \alpha_2 < \ldots < \alpha_{C-1}$; $\mathbf{x}_{ij}$ is a vector of covariates with vector of coefficients $\boldsymbol{\beta}$; the $u_j$s are random effects accounting for the correlation in the responses of the level 1 units belonging to the same level 2 unit.

Usually, to get identification, the overall intercept is omitted, thus $u_j$ can be regarded as a random shift of the intercepts so that the set of intercepts of the $j$th level 2 unit is $\alpha_c - u_j$, $c = 1, 2, \ldots, C-1$. The standard assumption on the $u_j$s is that, conditionally on the covariates, they are independent and normally distributed with zero mean and cluster variance $\sigma_u^2$.

Empirical Bayes prediction is the most widely used method for assigning values to random effects [14], by considering the means or the modes (EB modal predictions) of the empirical posterior distribution of the $u_j$s. Apart from linear models, numerical integration methods are needed to obtain the means of the posterior distribution, while EB modal predictions are obtained using gradient methods which do not require numerical integration. The assumed (prior) distribution of random effects has little impact on the mean square error of EB predictions, while it may severely affect the shape of the distribution of EB predictions, which tends to reflect the assumed one [10, 11]. However, when level 2 units have large sample sizes the predictions are highly reliable, in this case, the shape of the distribution of EB predictions could be informative about the true underlying shape of the random effects distribution, especially when also the number of level 2 units is high.

We exploit the EB modal predictions to cluster level 2 units by means of the density-based approach of [5] and [9], as implemented in the R package `pdfCluster` [4]. In this approach, clusters are detected as sets of sample points around the modes of a nonparametric estimate of the underlying density. Although any nonparametric density estimator could be used for this purpose, [5] and [9] consider the popular kernel density estimator (see, for example, [16]). The package `pdfCluster` allows for the use of different kernel functions and different strategies to tackle the task of bandwidth selection for kernel estimation [4]. In the next section, we illustrate the clustering approach outlined above through an application to hierarchical data with level 2 units having large sample sizes.

## 3 Application

The European Social Survey (ESS) is an academically driven cross-national survey that has been conducted every 2 years across Europe since 2001.[1] In this application, we use data from the fifth round (2010), with 50781 individuals nested into 26 countries. The number of individuals per country ranges from 1083 (Cyprus) to 3031 (Germany).

Our aim is to cluster the participating countries with respect to trust in the electoral system. Political trust can be explained by a combination of individual, cultural, and institutional factors. To this end, we consider *trust in parliament*, an ordinal variable measured on a scale from 0 (not trust at all) to 10 (complete trust). Figure 1 reports the overall distribution *trust in parliament* and box plots by country. The overall distribution is asymmetric with a peak in zero; this feature suggests to use an ordinal response model. The box plots show substantial variation both within and between countries.

Countries with low political trust (median $\leq 2$) are: Bulgaria (BG), Greece (GR), Croatia (HR), and Ukraine (UA). While countries with high political trust

[1] http://www.europeansocialsurvey.org/.

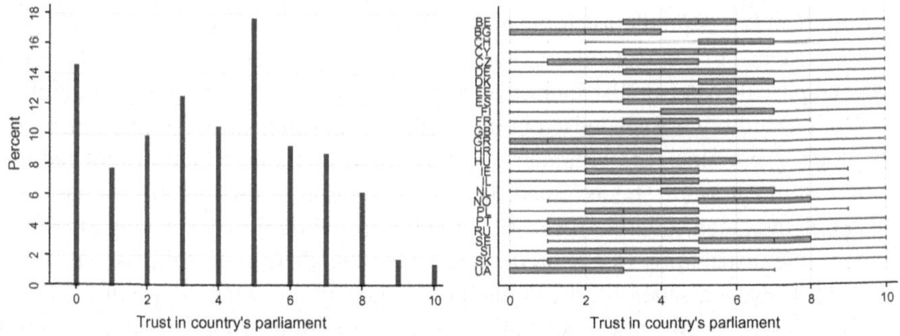


**Fig. 1** *Trust in parliament*: overall distribution (left panel), box plots by country (right panel)

(median $\geq 6$) are: Switzerland (CH), Denmark (DK), Finland (FI), the Netherlands (NL), Norway (NO), and Sweden (SE).

We aim to identify clusters of countries with similar levels of trust in parliament, adjusting for differences in subject's characteristics.

The data have a hierarchical structure, with citizens (level 1 units) nested within countries (level 2 units). We fit the random intercept model of Sect. 2 on *trust in parliament*, with $C = 11$ and $J = 26$. The random effects $u_j$ represent unobserved factors at the country level, interpreted as country's parliament trustworthiness. Model fitting is performed with the R package `ordinal` which yields ML estimates using adaptive Gaussian quadrature (Christensen, 2010). The `ordinal` package returns EB modal predictions of the random effects $u_j$. The predicted random effects are used to cluster the countries on the basis of the density-based algorithm outlined in Sect. 2, using the package `pdfCluster` [4].

## 3.1 Null Model

We start the analysis by fitting the null model, i.e., model (1) without covariates. The estimated cluster variance is $\widehat{\sigma}_u^2 = 0.848$, which is highly significant, thus confirming a relevant between-country variability.

The left panel of Fig. 2 reports the usual caterpillar plot of the EB modal predictions of random effects. This plot represents each country prediction alongside with the 95% interval. In this application intervals are narrow due to the large number of subjects per country, thus the predictions have a high reliability. The caterpillar plot does not entail a criterion for classifying countries, thus it is useful to apply a clustering procedure like the density-based method implemented in `pdfCluster`.

The right panel of Fig. 2 represents a kernel density estimate of the distribution of the EB modal predictions, obtained using a Gaussian kernel and the Normal reference rule (which is the default method of `pdfCluster`) for bandwidth selection (different strategies for bandwidth selection have been tried, obtaining the same countries

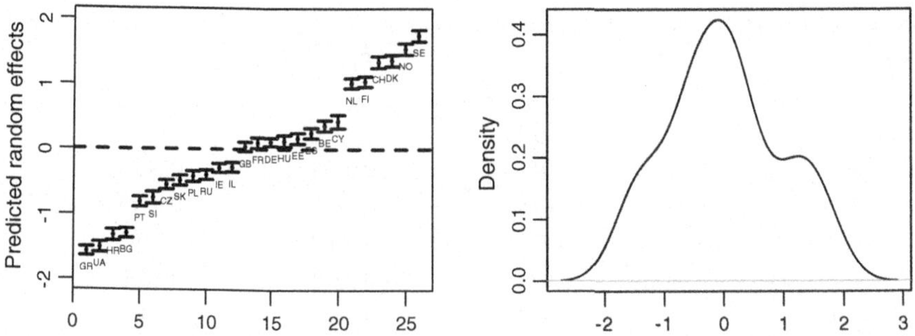


**Fig. 2** EB modal predictions of random effects from model (1) without covariates: caterpillar plot (left panel), kernel density estimate (right panel)

classification). The estimated density shows two modes, indeed the `pdfCluster` algorithm detects two clusters: one including all and only the countries with a median trust in parliament greater than 6, i.e. Switzerland (CH), Denmark (DK), Finland (FI), the Netherlands (NL), Norway (NO), and Sweden (SE).

## 3.2 *Model with Covariates*

In order to adjust differences across countries in terms of demographic and socioeconomic factors, we add to the random intercept model the following variables, measured at the subject level: *age* (centered at 48), *female* (1 yes, 0 no), *years of education*, *coping well with present income* (1 yes, 0 no). We consider also the country-level variable HDI (Human Development Index,[2] constructed using objective data on GDP per capita, education, and estimated life expectancy at birth). Among the considered countries, HDI ranges from 0.79 (Ukraine) to 0.97 (Norway), with an average of 0.92.

The estimated regression coefficients $\widehat{\boldsymbol{\beta}}$ of model (1) cannot be directly interpreted in terms of covariate effects on the probabilities of interest. Therefore, in Table 1 we report the predicted probabilities for $Y > 6$, namely the probability of high trustworthiness. The baseline subject of Table 1 is defined as follows: male, aged 48, no university- level education, not coping well on present income, living in a country with HDI equal to 0.79.

The estimated country-level standard deviation $\widehat{\sigma}_u = 0.4496$ is high, thus the country trustworthiness $u_j$ has a great effect on the predicted probabilities. For a baseline subject living in a country with an average value of the random effect, i.e., $u_j = 0$, the predicted probability of high trustworthiness is equal to 0.0376. This probability becomes about one half (0.0157) if we consider a baseline subject living in a country with $u_j = -2\widehat{\sigma}_u$, while it doubles (0.0877) in the case of $u_j = +2\widehat{\sigma}_u$.

[2]Details on HDI are at http://essedunet.nsd.uib.no/cms/topics/multilevel/ch6/all.html.

**Table 1** Full model results: predicted probabilities for $Y > 6$

| Subject | $\hat{P}(Y_{ij} > 6 \mid \mathbf{x}_{ij}, u_j)$ | | |
|---|---|---|---|
| | $u_j = -2\widehat{\sigma}_u$ | $u_j = 0$ | $u_j = +2\widehat{\sigma}_u$ |
| *baseline*[a] | 0.0157 | 0.0376 | 0.0877 |
| Female | 0.0147 | 0.0354 | 0.0828 |
| Age = 58 | 0.0157 | 0.0377 | 0.0879 |
| University-level education | 0.0213 | 0.0508 | 0.1162 |
| Coping well on present income | 0.0237 | 0.0564 | 0.1280 |
| HDI = 0.89 | 0.0380 | 0.0884 | 0.1925 |

[a]*baseline*: male, age = 48, no university-level education, not coping well on present income, HDI = 0.79

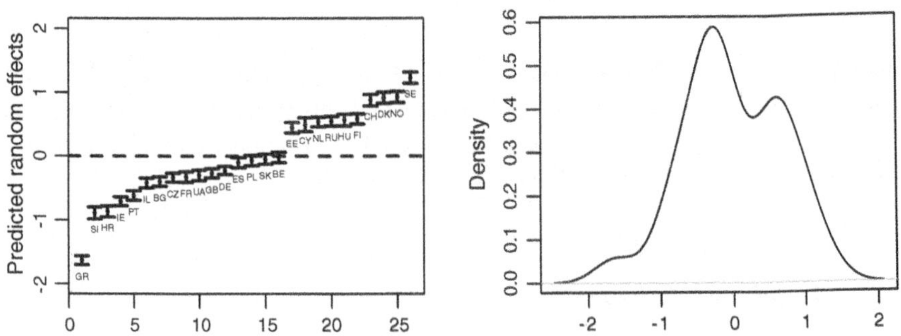


**Fig. 3** EB modal predictions of random effects from model (1): caterpillar plot (left panel), kernel density estimate (right panel)

All the covariates, but female, have a positive effect on the probability of high trustworthiness, for any value of the random effect. For example, considering $u_j = 0$, the probability of high trustworthiness is 0.0376 for the baseline subject, and it rises to 0.0508 if the subject has a university-level education.

Figure 3 reports the caterpillar plot of the EB modal predictions and the corresponding kernel density estimate. It is worth to note that Greece (GR) shows the lowest EB modal prediction.

We repeat the density-based clustering using the EB modal predictions from the model with covariates. Using a Gaussian kernel with the bandwidth selected according to the Normal reference rule, the procedure identifies the following two clusters:

1. *Medium–Low Trustworthiness* cluster: Belgium (BE), Bulgaria (BG), Czech Republic (CZ), Germany (DE), Spain (ES), France (FR), United Kingdom (GB), Greece (GR), Croatia (HR), Ireland (IE), Israel (IL), Poland (PL), Portugal (PT), Slovenia (SI), Slovakia (SK), and Ukraine (UA);
2. *High Trustworthiness* cluster: Switzerland (CH), Cyprus (CY), Denmark (DK), Estonia (EE), Finland (FI), Hungary (HU), the Netherlands (NL), Norway (NO), Sweden (SE), and Russian Federation (RU).

Thus, adjusting for covariates, 4 countries from the null model *medium–low trustworthiness* cluster joined the *high trustworthiness* cluster: Cyprus (CY), Estonia (EE), Hungary (HU), and Russian Federation (RU).

## 4 Final Remarks

In this work, we apply an unsupervised approach for clustering level 2 units in a multilevel setting exploiting the idea of density-based clustering, where the clusters are defined as high-density regions in the data space.

We apply this approach to ESS data as an exploratory tool to detect clusters of countries using a two-level model for ordinal responses.

The proposed approach deserves further investigation, along with comparisons with other methods for clustering level 2 units such as the latent class multilevel model [15], which relies on the assumption of a discrete distribution for random effects.

## References

1. Aitkin, M.: A general maximum likelihood analysis of variance components in generalized linear models. Biometrics **55**, 117–128 (1999)
2. Azzimonti, L., Ieva, F., Paganoni, A.M.: Nonlinear nonparametric mixed-effects models for unsupervised classification. Comput. Stat. **28**, 1549–1570 (2013)
3. Asparouhov, T., Muthén, B.: Multilevel mixture models. In: Hancock, G.R. and Samuelsen, K.M. (Eds.) Advances in latent variable mixture models, pp. 27–51, Charlotte, NC: Infurmation Age Publishing (2008)
4. Azzalini, A., Menardi, G.: Clustering via nonparametric density estimation: the R Package pdfCluster. J. Stat. Software **57**, 1–26 (2014)
5. Azzalini, A., Torelli, N.: Clustering via nonparametric density estimation. Statist. Computing **17**, 71–80 (2007)
6. Grilli, L., Rampichini, C.: Multilevel models for ordinal data. In: Kenett, R. and Salini, S. (eds.) Modern Analysis of Customer Surveys: with Applications using R., Chapter 19, Wiley (2012)
7. Heinzl, F., Tutz, G.: Clustering in linear mixed models with approximate Dirichlet process mixtures using EM algorithm. Stat. Modell. **13**, 41–67 (2013)
8. Lukociene, O., Varriale, R., Vermunt, J.K.: The simultaneous decision(s) about the number of lower- and higher-level classes in multilevel latent class analysis. Sociological Meth. **40**, 247–283 (2010)
9. Menardi, G., Azzalini, A.: An advancement in clustering via nonparametric density estimation. Statist. Computing **124**, 753–767 (2014)
10. McCulloch, C.E., Neuhaus, J.M.: Prediction of random effects in linear and generalized linear models under model misspecification. Biometrics **67**, 270–279 (2011)
11. McCulloch, M.E., Neuhaus, J.M.: Misspecifying the shape of a random effects distribution: why getting it wrong may not matter. Stat. Sci. **26**, 388–402 (2011)
12. Nylund, K.L., Asparouhov, T., Muthén, B.O.: Deciding the number of classes in latent class analysis and growth mixture modeling: a Monte Carlo simulation study. Struct. Equat. Model. **14**, 535–569 (2007)

13. Ohlssen, D.I., Sharples, L.D., Spiegelhalter, D.J.: Flexible random-effects models using Bayesian semi-parametric models: applications to institutional comparisons. Statist. Med. **26**, 2088–2112 (2007)
14. Skrondal, A., Rabe-Hesketh, S.: Prediction in multilevel generalized linear models. J. R. Stat. Soc. Ser. A **172**, 659–687 (2009)
15. Vermunt, J.K.: Mixture models for multilevel data sets. In: Hox, J., Roberts, J.K. (eds.) Handbook of Advanced Multilevel Analysis, pp. 59–81. Routledge (2010)
16. Wand, M.P., Jones, M.C.: Kernel smoothing. Chapman & Hall, London (1995)

# Clustering Macroseismic Fields by Statistical Data Depth Functions

**Claudio Agostinelli, Renata Rotondi and Elisa Varini**

**Abstract** The macroseismic intensity is an ordinal variable that describes the seismic damage effects of an earthquake. The collection of intensity values recorded at sites of an area hit by an earthquake is called macroseismic field; it constitutes the only information on historical earthquakes for which no instrumental recordings are available. Around the area of the epicenter, lines bounding points of equal seismic intensity (isoseismal lines) are used to represent the spatial distribution of the macroseismic intensities, and their shapes can suggest the characteristics of the earthquake source. Our aim is to identify clusters of macroseismic fields according to the size and shape of their isoseismal lines, or in seismological terms, according to the trend of macroseismic attenuation. First, the isoseismal lines of some fields are approximated by convex hulls. Then, fixed an intensity value, the set of the corresponding convex hulls are analyzed on the basis of statistical data depth functions. This nonparametric method ranks the convex hulls according to their statistical depth values, which in the present study are defined by the modified local half-region depth function. A similarity measure based on the same depth function allows us to compare all pairs of hulls and build a dissimilarity matrix to which a clustering procedure is applied in order to detect clusters of fields homogenoeus from the attenuation viewpoint. This method is illustrated on both simulated and real macroseismic data.

**Keywords** Clustering · Isoseismal lines · Pattern recognition · Similarity · Spatial data analysis

---

C. Agostinelli
Dipartimento di Matematica, Università di Trento, Trento, Italy
e-mail: claudio.agostinelli@unitn.it

R. Rotondi (✉) · E. Varini
CNR—Istituto di Matematica Applicata e Tecnologie Informatiche Enrico Magenes, Milano, Italy
e-mail: reni@mi.imati.cnr.it

E. Varini
e-mail: elisa@mi.imati.cnr.it



F. Mola et al. (eds.), *Classification, (Big) Data Analysis and Statistical Learning*, Studies in Classification, Data Analysis, and Knowledge Organization,
https://doi.org/10.1007/978-3-319-55708-3_16

# 1 Introduction

The severity of an earthquake is nowadays described by the magnitude, which indirectly measures the amount of energy released by the shock through the instrumental records (seismograms) produced by a seismograph. Reliable measurements of magnitudes are available from the 1960s.

On the contrary, the sizes of historical events are given in terms of the macroseismic intensity, which is an ordinal variable that is expressed through different scales; hereinafter we refer to the modified Mercalli–Cancani–Sieberg scale (MCS) [1]. Intensities are usually expressed in Roman numerals, whereby most scales have 12 degrees of intensity, with values that increase with severity. These intensities are more closely related to the effects produced by an earthquake on humans, buildings, and the natural environment. The collection of macroseismic intensities recorded at sites in an area surrounding a seismic source constitutes the macroseismic field of an earthquake. The graphical representation of these values on a map provides rapid information on the width of the area affected by the earthquake shaking. Then areas affected by equivalent levels of damage are identified by using noncrossing contours, in such a way as to enclose as many places as possible that have a macroseismic intensity that exceeds a fixed value, while keeping these curves smooth. The resulting lines are known as isoseismals (i.e., lines of equal shaking). Such maps are also useful, first, to determine what sort of effects might be expected from earthquakes in the future, once how rapidly the shaking decreases with distance has been estimated (i.e., the attenuation), and second, to define other information relating to the earthquake, such as the position of the epicenter, which can be approximated by the center of the innermost isoseismal. Moreover, attenuation has an important role in seismic hazard assessment and in the actions taken for risk reduction.

A macroseismic field is affected by several factors, such as the type of seismic source, the geological-tectonic setting, and the type of urban settlements and their behavior under seismic load. To forecast the effects of a future earthquake, a beta-binomial model that takes into consideration both circular and elliptical decay trends was proposed [2]. In the framework of the European project "Urban Prevention Strategies using Macroseismic and Fault sources" (2012–2013, Grant Agreement n. 230301/2011/613486/SUB/A5), this model was applied to the seismicity of some European countries. Furthermore, a method was also investigated to identify the common pattern of isoseismal lines in a set of macroseismic fields through the use of statistical data depth functions [3]. It is clear that the more homogeneous a dataset is, the more reliable the result is. Hence, in this chapter, our aim is to define a method to identify clusters of macroseismic fields according to the sizes and shapes of their isoseismal lines, or in other words, according to their macroseismic attenuation trends.

## 2 Statistical Tools: Depth Functions and Similarity Measures

The macroseismic field of an earthquake is a set of geographical coordinates (i.e., latitude, longitude) that are each associated with an intensity value $I_s$ that indicates how strongly the earthquake hit that site (Fig. 1a).

Macroseismic fields may be expressed as functional datasets in this way. Let $\Delta I = I_0 - I_s$ denote the intensity decay at a site, where $I_0$ is the intensity at the epicenter of the earthquake. First, the geographical coordinates of each site are converted into Cartesian coordinates with the origin at the epicenter. Then, for any value of the intensity decay $\Delta I$, a convex hull is built that encloses all of the sites where an intensity decay not exceeding $\Delta I$ was observed (Fig. 1b). Finally, the vertices of the convex hull are converted from Cartesian to polar coordinates, where the first is the angular coordinate measured anticlockwise from the positive horizontal axis, and the second corresponds to the site-epicenter distance. This leads to a piecewise linear function, as shown in Fig. 1c. Changing the sense of rotation corresponds to reverse the function with respect to the vertical axis, while changing the zero direction means

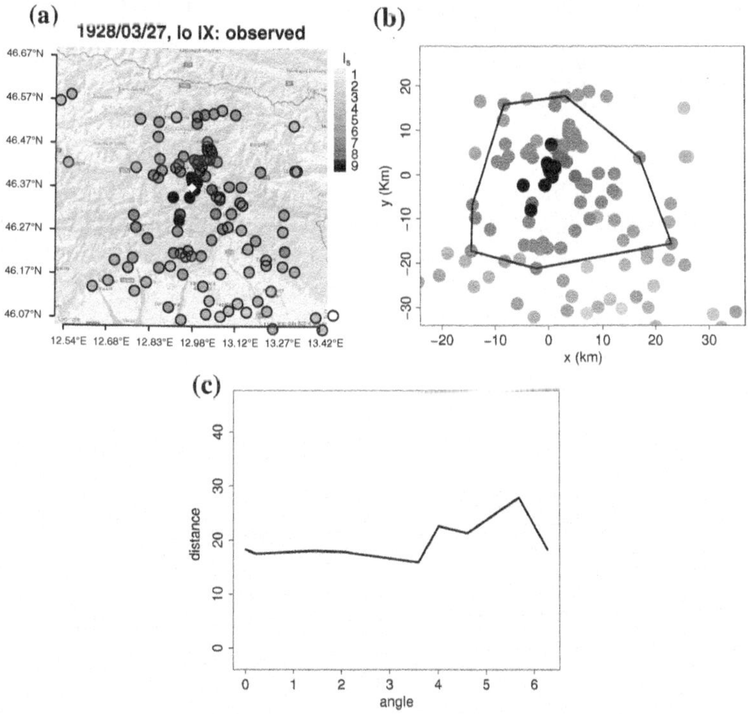


**Fig. 1** **a** Macroseismic field of an earthquake with epicentral intensity $I_0 = IX$. **b** Convex hull obtained for $\Delta I = 2$. **c** Piecewise linear function

to translate the function along the horizontal axis. Both these changes do not have any effect on the analysis we perform.

By repeating the procedure for every macroseismic field of the seismic region that is considered, we collect some functional datasets, as one for each value of the intensity decay $\Delta I$ (Fig. 2).

By analyzing each functional dataset by a data depth function, it is possible to identify the most central hull that represents the attenuation pattern. Data depth is a nonparametric method that was designed to order points of a multivariate space from their center outwards. Recently, this method has been extended to the ordering of functions and trajectories [4, 5].

Many different definitions of depth are currently used (for a review, see [6], and references therein). Among these, the local version of the modified half-region depth is appropriate for the analysis of irregular (nonsmooth) curves with many crossing points, as for the convex hulls here [4].

Let $C(T)$ be the space of the real continuous functions on some compact interval $T \subset \Re$. Given $y \in C(T)$, the hypograph (epigraph) of $y$ is defined as the set of points lying on or below (above) its graph. The modified half-region depth $d_{MHR}(y; \mathbf{y}_n)$ of $y$ with respect to a collection of functions $\mathbf{y}_n = \{y_j \in C(T) : j = 1, \ldots, n\}$ is equal to the minimum between the sample means of the proportions of the segment lengths in which the $y_i$ are in the epigraph and the hypograph of $y$, respectively:

$$d_{MHR}(y; \mathbf{y}_n) = min \left( \sum_{j=1}^{n} \frac{\lambda(t \in T : y(t) \leq y_j(t))}{n\lambda(T)}, \sum_{j=1}^{n} \frac{\lambda(t \in T : y(t) > y_j(t))}{n\lambda(T)} \right),$$

where $\lambda$ denotes the Lebesgue measure on $\Re$.

The local version $ld_{MHR}(y; \mathbf{y}_n, \tau)$ is obtained by reducing the hypograph of $y$ to the slab that is bounded by the two curves $y - \tau$ and $y$, for a constant value $\tau > 0$. Analogously, the epigraph of $y$ is the slab between $y$ and $y + \tau$. The depth of $y$ with respect to $\mathbf{y}_n$ turns out to be defined by

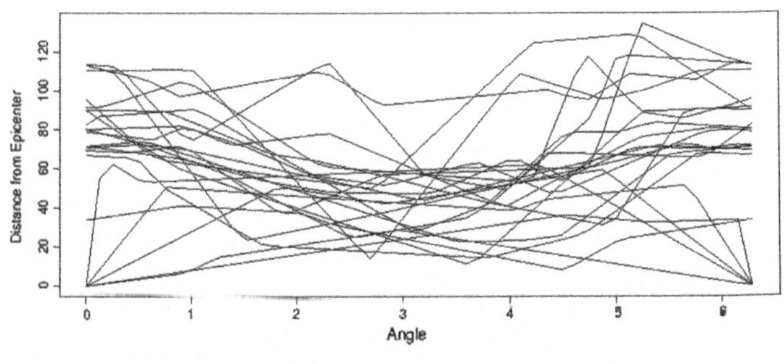


**Fig. 2** Functional dataset obtained from all of the convex hulls with a fixed intensity decay $\Delta I$

$$ld_{MHR}(y;\mathbf{y}_n,\tau) = min\left(\sum_{j=1}^{n}\frac{\lambda(t\in T : y(t)\le y_j(t))}{n\lambda(T)}\cdot \mathbb{1}_j(y,y;\tau),\right.$$
$$\left.\sum_{j=1}^{n}\frac{\lambda(t\in T : y(t)> y_j(t))}{n\lambda(T)}\cdot \mathbb{1}_j(y,y;\tau)\right),$$

where $\mathbb{1}_j(z,w;\tau) = \mathbb{1}\left[z(t)-\tau\le y_j(t)\le w(t)+\tau;\ \forall t\in T\right]$, for any $z, w \in C(T)$.

Let $y_{(1)}, y_{(2)}, \ldots, y_{(n)}$ denote the functions of the set $\mathbf{y}_n$, arranged in increasing order according to their depth values with respect to $\mathbf{y}_n$. A center-outward order is induced on these curves, so that $y_{(1)}$ is the most outlying curve and $y_{(n)}$ is the most central curve.

If we think that there might be multiple centers, it would be appropriate to compare pairs of curves through a similarity measure based on the same depth function. The local modified half-region depth similarity $ls_{MHR}(x, y; \mathbf{y}_n, \tau)$ of the trajectory pair $(x, y)$ is defined by [7]

$$ls_{MHR}(x,y;\mathbf{y}_n,\tau) = min\left(\sum_{j=1}^{n}\frac{\lambda(t\in T : z(t)\le y_j(t)\le z(t)+\tau)}{n\lambda(T)}\cdot \mathbb{1}_j(z,w;\tau)\right. \quad (1)$$
$$\left.\sum_{j=1}^{n}\frac{\lambda(t\in T : w(t)-\tau\le y_j(t)\le w(t))}{n\lambda(T)}\cdot \mathbb{1}_j(z,w;\tau)\right)$$

where
$$z = x\vee y = \{(t,z(t)) : t\in T, z(t) = \max[x(t),y(t)]\},$$
$$w = x\wedge y = \{(t,w(t)) : t\in T, w(t) = \min[x(t),y(t)]\}.$$

The adopted similarity measure is such that for all $x, y \in C(T)$, $ls_{MHR}(x, x; \mathbf{y}_n, \tau) = ld_{MHR}(x; \mathbf{y}_n, \tau)$, and

$$ls_{MHR}(x,y;\mathbf{y}_n,\tau) \le min\left(ld_{MHR}(x;\mathbf{y}_n,\tau), ld_{MHR}(y;\mathbf{y}_n,\tau)\right). \quad (2)$$

Then, this similarity measure is used to construct the following dissimilarity matrix $\mathbf{D} = \left(D_{ij}\right)_{i,j=1}^{n}$ for $\mathbf{y}_n$ [8]:

$$D_{ij} = \sqrt{ls_{MHR}(y_i,y_i;\mathbf{y}_n,\tau) + ls_{MHR}(y_j,y_j;\mathbf{y}_n,\tau) - 2\cdot ls_{MHR}(y_i,y_j;\mathbf{y}_n,\tau)}. \quad (3)$$

Applying hierarchical cluster analysis based on different methods to the dissimilarity matrix $\mathbf{D}$, we group the original macroseismic fields in clusters according to the shape of the generated convex hulls. This analysis is performed through the algorithm *hclust*, implemented by the free software R (www.R-project.org, [9]).

## 3 Application to Simulated and Real Datasets

To select clustering methods that are more suitable to our problem, we first analyzed some sets of simulated fields (in Sect. 3.1 we present the results concerning two of these sets), and then a real dataset (Sect. 3.2).

### 3.1 Analysis of Simulated Datasets

We simulated two sets of 60 fields with epicentral intensity $I_0 = VIII$, each composed of $N = 100$ sites with an associated felt intensity. Each set is divided into three groups where the simulated spatial distribution of the felt intensity points is such that the respective isoseismal lines differ in shape (i.e., circular, elliptical), size, $\epsilon$ eccentricity, and $\phi$ rotation angle. In the first set, group A is formed by 24 fields that are characterized by circular isoseismal lines with radius of the first isoseismal $r = 3$; groups B and C are each formed by 18 fields where the isoseismal lines are ellipses with semi-minor axis $b = 3$ and semi-major axis $a = 2\,b$ of the first isoseismal (i.e., $\epsilon = 0.866$), and with rotation angles $\phi = 0$ and $\phi = \pi/4$, respectively. The second set differs from the first one just for the size of the semi-minor axis which is $b_B = 4$ and $b_C = 5$ in group B and C respectively; the eccentricity remains the same. Without loss of generality, the epicenter of each field is located at $(0, 0)$.

The simulation algorithm of a field can be formalized by repeating the following procedure from $i = 1$ to $N$:

1. Draw a polar angle $\theta_i$ from a uniform distribution on $[0, 2\pi)$,
2. If the shape of the attenuation decay is
   - circular: find the Cartesian point $(x_i, y_i)$ that corresponds to $\theta_i$ on a circle with center $(0, 0)$ and radius $r$;
   - elliptical: find the Cartesian point $(x_i, y_i)$ that corresponds to $\theta_i$ on an ellipse with center $(0, 0)$, length of semi-major axis $a$, length of the semi-minor axis $b$, and rotation angle $\phi$,
3. Set the felt intensity $I_i = \left[2\left(I_0 - I\frac{I_0-1}{max(I)} - 0.5\right)\right]/2$, where $I$ is a sample from an exponential distribution with rate 2,
4. Apply a Gaussian perturbation to the Cartesian point: $x_i = px_i,\ y_i = py_i$, where $p \sim N(r(I_0 - I_i), 1)$.

The various methods of hierarchical clustering we have considered are characterized by the definition of inter-cluster distance and mainly belong to two classes: methods which tend to produce long thin clusters, such as single-linkage, centroid, median method, and methods which tend to find compact spherical clusters, such as complete, average, McQuitty's, and Ward's method [10]. Applying the methods of the first class, the so-called chaining phenomenon occurs so that most of the fields turns out to belong to the same cluster. Better results are produced by the methods of the other class; in particular all methods distinguish fields with circular isoseismal lines from those with elliptical isoseismal lines. In the second set of simulated fields,

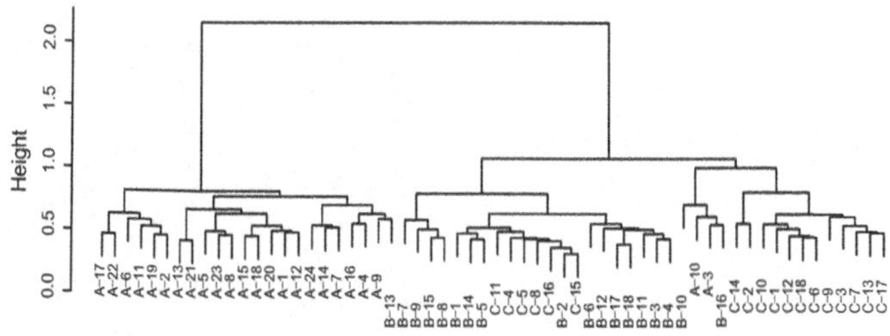


**Fig. 3** Dendrogram of the first dataset showing the three groups of simulated fields: A, B, C

the methods identify properly groups B and C whereas in the first set just Ward's method is able to distinguish the rotated fields by $\phi = \pi/4$ from the other elliptical fields.

Figure 3 shows the dendrogram that results from the hierarchical cluster analysis—by Ward's method—of the first simulated dataset. The three simulated groups are essentially well identified: the cluster on left is composed by 22 fields of group A and 1 of group B, the central cluster is formed by 15 fields of group B and 6 of group C, while the cluster on right contains 2 fields of group A, 2 of group B, and 12 of group C. This slight misclassification may be due to the fact that the simulation of B and C differs only for the relatively small rotation angle.

The same results were also achieved by adopting the nonhierarchical clustering method *k-means*. To do that we first applied a multidimensional scaling technique [8] for embedding dissimilarity information in two-dimensional Euclidean space by representing the set of the convex hulls as a set of points in such a way that the Euclidean distances between points approximate the dissimilarities between the corresponding curves.

## 3.2 *Analysis of a Set of Italian Macroseismic Fields*

We consider here the macroseismic fields associated with 31 earthquakes of $I_0 = IX$ and $I_0 = IX - X$ of the Mercalli–Cancani–Sieberg scale, drawn from the DBMI11 Italian Macroseismic Database. The cluster analysis of this dataset produces the dendrogram in Fig. 4 (left), which shows a clear evidence of at least two main clusters. On the map in Fig. 4 (right), the epicenters of the earthquakes belonging to these two clusters are depicted by stars (left cluster) and circles (right cluster), respectively. Following the empirical rule that suggests cutting the dendrogram at the level where a jump in levels of two consecutive modes is large, we could again lower the threshold of the height value so as to obtain three clusters by subdividing the right cluster into two sets of fields whose epicenters are depicted by gray and black circles in Fig. 4 (right).

We note that no isoattenuation zone is clearly identifiable; on the contrary, fields of different clusters are placed even very close to each other. For example, let us consider the May 6, 1976 and March 27, 1928 earthquakes denoted by 1 and 2 in

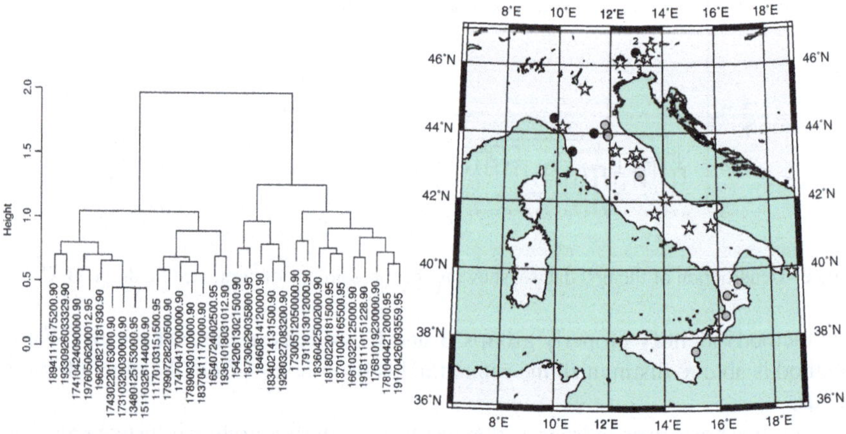


**Fig. 4** (Left) Dendrogram of the cluster analysis of the 31 Italian earthquakes with $I_0 = IX$ or $IX - X$. The individual events are labeled according to their date and epicentral intensity (separated by a dot). (Right) Epicenters of the 31 Italian earthquakes partitioned into the two main clusters (stars, circles) and the three sub-clusters (stars, black circles, gray circles)

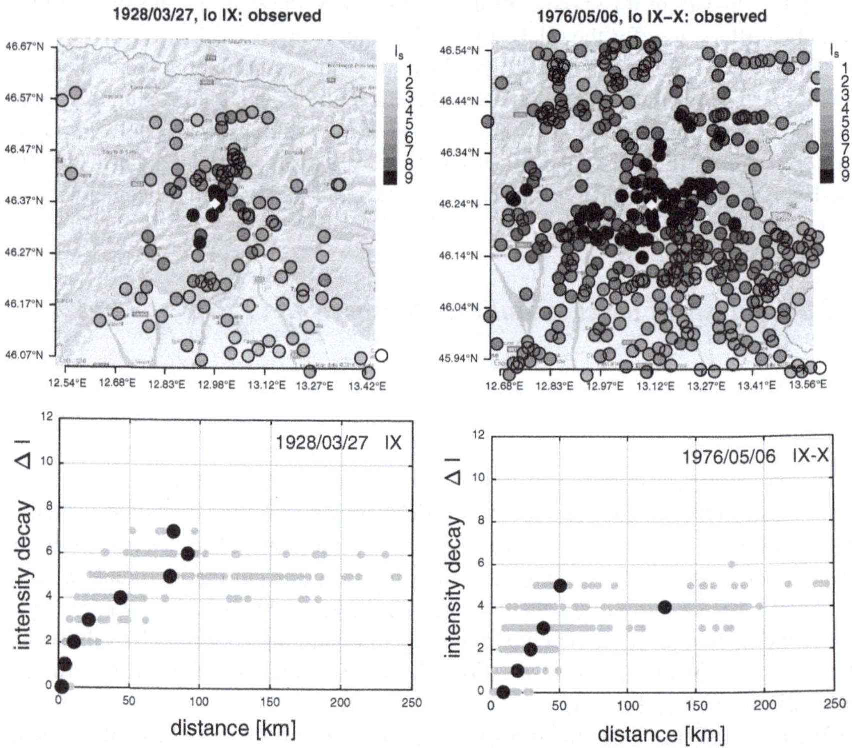


**Fig. 5** Representations of the 03/27/1928 earthquake (left) and the 05/06/1976 earthquake (right). Top: Macroseismic field of the earthquakes. Bottom: Plot of intensity decay versus epicentral distance, where the black dots denote the median of the distance subsets

Fig. 5 and located in the northeast of Italy. The macroseismic fields of these two events are shown in top panels of Fig. 5, while their attenuation trend is represented graphically in bottom panels where the gray dots at the same horizontal level denote the distances between epicenter and sites of the same decay $\Delta I$, and the black dot indicates the median of this set of distances. This visual inspection agrees with the membership of the two earthquakes in different clusters. Moreover, these quakes are associated with seismogenic sources characterized by faults with different prevalent rupture mechanism: left-lateral strike-slip fault as for the March 27, 1928 shock and reverse fault as for the May 6, 1976 shock [11]. Since, according to recent studies in seismology, seismic attenuation depends on the features of the seismic source, also this remark supports the result of our clustering procedure.

Since the seismic sources of earthquakes that occurred in historical periods are unknown, the clustering procedure we have proposed could be also exploited to extract information on these sources from the analysis of the attenuation trend of the corresponding macroseismic fields.

**Acknowledgements** The authors thank the two anonymous reviewers for their constructive comments and suggestions. Some maps were produced with GMT software [12].

## References

1. Sieberg, A.: Geologie der Erdbeben. Handbuch der Geophysik **2**(4), 552–555 (1930)
2. Zonno, G., Rotondi, R., Brambilla, C.: Mining Macroseismic Fields to Estimate the Probability Distribution of the Intensity at Site. Bull. Seismol. Soc. Am. **98**(5), 2876–2892 (2009)
3. Agostinelli, C., Rotondi, R.: Analysis of macroseismic fields using statistical data depth functions. Bull. Earthq. Eng **14**(7) (2016)
4. Agostinelli, C., Romanazzi, M.: Ordering curves by data depth. In: Giudici, P., Ingrassia, S., Vichi, M. (eds.) Statistical Models for Data Analysis, pp. 1–8, Springer (2013)
5. López-Pintado, S., Romo, J.: A half-region depth for functional data. Comput. Statist. Data Anal. **55**, 1679–1695 (2011)
6. Liu, R.Y., Serfling, R.J., Souvaine, D.L.: Data depth: robust multivariate analysis, computational geometry and applications. Series in Discrete Mathematics and Theoretical Computer Science, DIMACS **72** (2006)
7. Agostinelli, C.: Local half region depth for functional data. J. Multivariate Anal. **163**, 67–79 (2018)
8. Gower, J.C.: Some distance properties of latent root and vector methods used in multivariate analysis. Biometrika **53**, 325–338 (1966)
9. R Core team. R: A language and environment for statistical computing. R Foundation for Statistical Computing, Vienna, Austria (2017)
10. Kaufman, L., Rousseeuw, P.J.: Finding Groups in Data: An Introduction to Cluster Analysis, Wiley (2005)
11. Basili, R., Valensise, G., Vannoli, P., Burrato, P., Fracassi, U., Mariano, S., Tiberti, M.M., Boschi, E.: The Database of Individual Seismogenic Sources (DISS), version 3: summarizing 20 years of research on Italy's earthquake geology. Tectonophysics **453**(1–4), 20–43 (2008)
12. Wessel, P., Smith, W.H.F.: New, improved version of Generic Mapping Tools released. Eos Trans. AGU **79**(47), 579 (1998)

# Comparison of Cluster Analysis Approaches for Binary Data

**Giulia Contu and Luca Frigau**

**Abstract** Cluster methods allow to partition observations into homogeneous groups. Standard cluster analysis approaches consider the variables used to partition observations as continuous. In this work, we deal with the particular case all variables are binary. We focused on two specific methods that can handle binary data: the monothetic analysis and the model-based co-clustering. The aim is to compare the outputs performing these two methods on a common dataset, and figure out how they differ. The dataset on which the two methods are performed is a UNESCO dataset made up of 58 binary variables concerning the ability of UNESCO management to use Internet to promote world heritage sites.

**Keywords** Cluster analysis · Binary data · Monothetic analysis cluster · Model-based co-clustering · UNESCO

## 1 Introduction

The United Nations Educational, Scientific and Cultural Organization (UNESCO) preserves culture and naturalistic heritage sites located in all the 195 member countries. It keeps a list of world heritage sites, where the most important and beautiful sites in the world are included. This list includes different types of resources [10]: cultural heritage, natural heritage, mixed cultural and natural heritage, cultural landscapes, and movable heritage.

Culture and environment represent an important force to attract tourists and, in the last years, recent demographic, social, and cultural changes determined a rise of cultural tourism flows [4, 8, 9]. This kind of tourism is important for different reasons. For instance, it has a positive economic and social impact, and can help

G. Contu (✉) · L. Frigau
Department of Economics and Business, University of Cagliari,
Viale Sant'Ignazio 17, 09123 Cagliari, Italy
e-mail: giulia.contu@unica.it

L. Frigau
e-mail: frigau@unica.it

 
F. Mola et al. (eds.), *Classification, (Big) Data Analysis and Statistical Learning*,
Studies in Classification, Data Analysis, and Knowledge Organization,
https://doi.org/10.1007/978-3-319-55708-3_17

to change the seasonal tourism flows, to reinforce cultural identity, and to support preserving of heritage [9].

In order to obtain an adequate development of cultural tourism, it is necessary to consider two different elements: The first element is linked with economic and social development, and the second is the communication. Promotion and communication should be efficient and incisive. Visitors should find all information, also tourism information, before coming to a cultural destination. They should also be supported in the organization of their tours, before, during, and after them. This is possible only if the management of cultural or environmental sites creates adequate information and chooses a correct communication media.

In this work, we considered a specific communication tool as the website, inasmuch World Wide Web assumes a critical and important role in tourism choices [1]. More in details, we analyzed the quantity and quality of information present in a website of UNESCO cultural and environmental sites. The websites have become the most important and essential places where information can be found, tours can be planned, and information can be shared [11]. For this reason, it is really important to create an adequate website where tourists can find all the information they need. Also, sharing information, photos, and video, before, during, and after the tourism experience improves the capacity to attract tourists and increases tourism flows. This sharing represents a way to involve tourists and to help them to create their own tours.

As the website is a virtual place where tourists can find information, in the same way it becomes an important tool for marketing and communication, a support for business to promote products and services and to generate revenue [1, 6, 11].

Over time, websites assumed an important role for promoting tourism destination and also cultural and natural places. "*However, not all websites are equally successful*" [1]. All websites are different, each one presents different elements and is constructed in a different way. The problem is to understand if a website is designed to attract visitors and provide information. It is important to evaluate if website can be classify in relation to specific characteristics, for instance informative or touristic website, and to understand if there are similar behaviors among websites.

For this reason, we have decided to analyze the websites of cultural and natural heritages of UNESCO located in three countries bordering in the Mediterranean Sea: France, Spain, and Italy, analyzing 137 UNESCO sites that represent the 25% of all-European heritage sites. When we have started to find the websites, we have discovered that not all heritage sites have a website and, at the same time, some heritage sites have more than one site. For this reason, we have analyzed 142 websites. Moreover, it is evident that in some cases information about heritage sites are included in official destination websites. This is a way to consider and promote the heritage site as a fundamental element of specific destination, which is an expression of a touristic policy. For this reason when dataset has been created, it has been checked if the website recorded was either an official destination website or specific website for the UNESCO heritage.

The dataset was composed by 65 websites of UNESCO heritage and 69 binary variables, concerning these characteristics of the website:

- contact and support in researching information: in this part, we analyzed both the presence of contacts and the usability of websites;
- adequacy of reported information, distinguished in four more parts that are finalized to understand if:
  - the brand of UNESCO or other brands are used to recognize the destination and the location;
  - the general information about the places and the tourism information are given in the website. The aim of this area is to figure out if the website is used as a tool to promote the destination and to support tourists in the organization of their tours;
  - in the websites, it is possible to find links to local tourist institution, tourism product club, and public institution. The goal is to understand if the cultural and natural heritage is considered as a part of the destination and if it is integrated in the network of tourism destinations;
  - news and articles from newspapers are included in the website so as to inform visitors of events and other activities realized in the cultural and natural sites.
- relational skills, in this area we wanted to understand the activities realized by websites manager to create a relation with virtual visitors;
- other information/internal communication. The aim of this part is to understand if in the websites data about the performances of the site (for instance: number of visitors), information about UNESCO list and planes for tourism development are also included; finally, a part reserved for tour operators.

In order to analyze this binary variables, we have decided to use two different cluster methods: MONA cluster and model-based co-clustering. We want to compare different results and identify the more useful method for the analysis of binary data.

The rest of the paper is organized as follows. Section 2 recalls the main features of the two cluster methods. Section 3 presents the results of their performances on UNESCO data and Sect. 4 ends the paper with some concluding remarks.

## 2 Methods

### 2.1 *Monothetic Analysis Cluster*

The monothetic analysis (MONA) is a hierarchical divisive cluster method used for binary variables [3, 5, 7]. At each step, the MONA algorithm splits the set of observations into two subsets, using a selected variable: in the first subset, all observations assume a value equal to zero for the selected variable, in the second subset equal to one . This means that in one subset, it is possible to identify the presence of the attribute, and in the other one the attribute is not present . This separation step is replayed until either each subset contains only one single object or no variables can be more used because all observations have the same value for each

**Table 1** Contingency table between variables $f$ and $g$

| | | Variable $g$ | |
|---|---|---|---|
| | | 1 | 0 |
| Variable $f$ | 1 | $a_{fg}$ | $b_{fg}$ |
| | 0 | $c_{fg}$ | $d_{fg}$ |

one. The variable $t$ selected for splitting a subset is the variable with the maximal total association to the other ones. In order to select that, the method based on pairwise measure of association for variables is used .

Let $f$ and $g$ be two variables, the association between them is measured with the following formula:

$$A_{fg} = |a_{fg}d_{fg} - b_{fg}c_{fg}| \tag{1}$$

where the values $a_{fg}$, $d_{fg}$, $b_{fg}$ and $c_{fg}$ are obtained from the Table 1. For each variable, $f$ is calculated a total measure $A_f$ summing the association measures between $f$ and the other variables

$$A_f = \sum_{g \neq f} A_{fg} \tag{2}$$

In order to select among all $f$ variables, the variable $t$ used to split the subset, the total measure $A_f$ is maximized

$$A_t = \max_f A_f \tag{3}$$

If more maximal values will be the same for two or more variables, then the one appearing first will be selected as $t$.

## 2.2 Model-Based Co-clustering

Model-based co-clustering is particular cluster analysis method that constructs an optimal partition considering simultaneously both objects and variables, and organizes them into a $g \times m$ homogeneous blocks, where $g$ and $m$ are, respectively, the numbers, defined *ex ante*, of row and column clusters. Several co-clustering models exist, such as for binary, contingency, continuous, and categorical data. Since in this work we deal with binary data, we focus on co-clustering binary model. For this kind of data, Bernoulli latent block models are used.
Let $\mathbf{x}$ denote a $n \times d$ data matrix defined by $\mathbf{x} = \{(x_{ij}); i \in I \text{ and } j \in J\}$, where $I$ is a set of $n$ objects (i.e., rows) and $J$ is a set of $d$ variables (i.e., columns). The parameters to estimate of those models are four: $\alpha$ the most frequent binary value of

the blocks, $\varepsilon$ the dispersion of the blocks (i.e., the probability of having a different value than the center), $\pi$ the row proportions, and $\rho$ the column proportions.
The parameter $\alpha$ is given by the matrix $\mathbf{p} = (p_{k\ell})$ where $p_{k\ell} \in [0, 1]$ $\forall k = 1, \ldots, g$ and $\ell = 1, \ldots, m$. The probability distribution $f_{k\ell}(x_{ij}, \mathbf{p}) = f(x_{ij}, p_{k\ell})$ is the Bernoulli distribution

$$f(x_{ij}, p_{k\ell}) = (p_{k\ell})^{x_{ij}}(1 - p_{k\ell})^{1-x_{ij}} \tag{4}$$

Re-parameterizing the model density [2], the parameters $p_{k\ell}$ of the Bernoulli mixture model are replaced by $\alpha_{k\ell}$ and $\varepsilon_{kj}$

$$f(x_{ij}, \mathbf{p}) = (\varepsilon_{kj})^{|x_{ij}-\alpha_{k\ell}|}(1 - \varepsilon_{kj})^{1-|x_{ij}-\alpha_{k\ell}|}, \tag{5}$$

where

$$\begin{cases} \alpha_{k\ell} = 0, \varepsilon_{k\ell} = p_{k\ell} & \text{if } p_{k\ell} < 0.5 \\ \alpha_{k\ell} = 1, \varepsilon_{k\ell} = 1 - p_{k\ell} & \text{if } p_{k\ell} > 0.5 \end{cases} \tag{6}$$

## 3 Comparison of the Methods on UNESCO Data

In the first phase, we applied the MONA algorithm. It splits the observations into seven steps as shown in Fig. 1. The MONA plot is a bar plot where the horizontal thick red bars indicate the variables selected to split, and the thin white lines between the bars indicate the observations. At each step, a partition is performed. If a red bar stops at a step, then the corresponding variable is selected for splitting the set into two subsets: all observations of the set above that bar assume a value equal to one for the selected variable and those below to zero. We considered the clusters obtained splitting the observations with two steps, because this partition allows to get significant clusters. In these two steps, the MONA algorithm selected the variable concerning "wine and food information" (CIU27) first, then the variables about "cart rental" (CIU20) and "accommodation" (CIU23). Analyzing these clusters, we understood that there are some websites characterized to give more useful information on how to contact the information office, to get to cultural and natural sites, and to visit the heritage sites. On the other hand, other websites are characterized to support the travel organization offering information about accommodation, means of transport, food and wine information, shopping, events, and touristic activities. The four clusters can be classified according to a range that changes from websites that are typically more touristic to websites that focus more on informative elements. However, it is not possible to label clusters as exclusively touristic or as exclusively informative, since at each step only one variable is used to split and consequently it is impossible to identify a clear distinction among them. It is also impossible to obtain a clear distinction considering the counting variables, because also in this case we can obtain roughly clusters with the same number of variables.

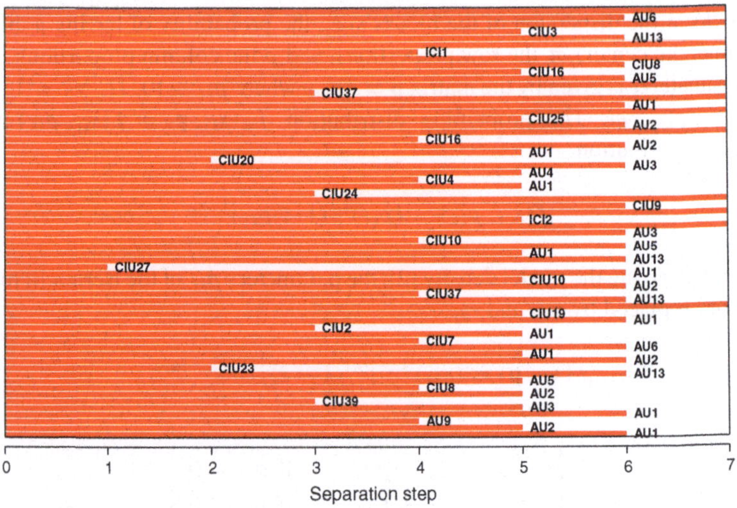


**Fig. 1** The MONA plot. In the zero step, we find a partition with just one cluster and all observations. In the step one, the observations are split in two subsets considering the variable "food and wine information" (CIU27). One subset is characterized by CIU27 equal to one, whereas the other one by CIU27 equal to zero. In the second step, when CIU27 is equal to one, the variable selected to split is "information about car rental" (CIU20), otherwise "list of accommodation available" (CIU23). Furthermore, in the step two, if both CIU27 and CIU20 are equal to one the variable selected to split is "link to other territories" (CIU37), if CIU27 is equal to one and CIU20 to zero, the variable selected is "availability for accommodation" (CIU24). The cluster identification gets on until the step seven

In the second phase, model-based co-clustering was applied. Since we needed four clusters of observations, we set the number of row clusters to four. Instead, the number of column clusters was set to three because it maximized the Integrated Complete Likelihood (ICL) at $-2254.109$. The output is shown in Fig. 2. The model-based co-clustering plot is made up of two parts: the original data and the co-clustered data. The original data is a matrix where the rows consist in observations and the columns in variables. The co-clustered data is the one obtained by performing the model-based co-clustering on the original data. It expresses the final output of the permutations of objects and variables made in order to minimize the dispersion level in each class, defined by the cross between a row cluster and a column cluster. As we set, we obtained 12 classes (4 row clusters $\times$ 3 column clusters). In order to compare this method and the MONA one, we considered just the partition obtained considering the observations, therefore the four row clusters, hence called simply "clusters".

Concerning the interpretation of the four clusters, in contrast to MONA clusters, model-based co-clustering defined groups in clear way. We can label the clusters considering the counting variables, and obtaining a range that changes from websites characterized by the presence of many variables to websites with less variables. Furthermore, we can define the clusters as totally touristic and totally informative.

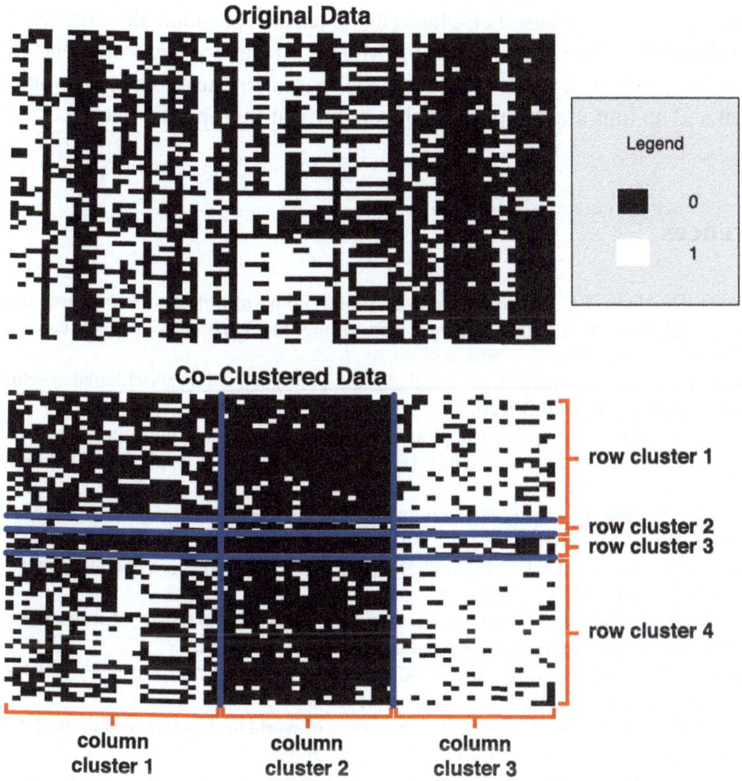


**Fig. 2** The model-based co-clustering plot

This happens because the observations are split considering all variables together, not only one at once as MONA algorithm.

## 4 Conclusions

We have applied two different cluster methods to the same dataset composed by binary variables to evaluate possible similarity and dissimilarity in term of results and plots. As said before, the first was the monothetic analysis (MONA) that is a hierarchical divisive cluster method. The second was the model-based co-clustering that is a particular cluster analysis method which considers simultaneously both objects and variables, and organizes them into a $g \times m$ homogeneous blocks.

Analyzing the results, we can say that the model-based co-clustering created homogeneous clusters with specific features, totally different from each other. In contrast, the MONA cluster created clusters more similar in terms of variables and size.

Furthermore, model-based co-clustering is used to partition all variables and all observation at the same time, instead the MONA cluster considers only one variable at a time. In conclusion, we can say that model-based co-clustering gives a significant partition and an important support in the analysis of this specific dataset.

## References

1. Bastida, U., Huan, T.C.: Performance evaluation of tourism websites' information quality of four global destination brands: Beijing, Hong Kong, Shanghai, and Taipei. J. Business Res. **67**(2), 167–170 (2014)
2. Bhatia, P., Iovleff, S., Govaert, G.: blockcluster: an r package for model based co-clustering. J. Stat. Software **76**(9), 1–24 (2017)
3. Greenwood, M.C.: A comparison of plots for monothetic clustering, with applications to microbial communities and educational test development. Electron. J. Appl. Stat. Anal. **5**(1), 1–14 (2012)
4. Ismail, N., Masron, T., Ahmad, A.: Cultural heritage tourism in malaysia: Issues and challenges. In: SHS Web of Conferences, vol. 12, p. 01059. EDP Sciences (2014)
5. Kaufman, L., Rousseeuw, P.J.: Finding groups in data: an introduction to cluster analysis, vol. 344. Wiley, USA (2009)
6. Law, R., Qi, S., Buhalis, D.: Progress in tourism management: a review of website evaluation in tourism research. Tourism management **31**(3), 297–313 (2010)
7. Maechler, M., Rousseeuw, P., Struyf, A., Hubert, M., Hornik, K.: cluster: Cluster analysis basics and extensions. R package version **1**(4) (2016)
8. Patuelli, R., Mussoni, M., Candela, G.: The effects of world heritage sites on domestic tourism: a spatial interaction model for Italy. J. Geographical Syst. **15**(3), 369–402 (2013)
9. Richards, G.: Production and consumption of european cultural tourism. Ann. Tourism Res. **23**(2), 261–283 (1996)
10. World Heritage Committee: Operational guidelines for the implementation of the World heritage Convention. Unesco World Heritage Centre (2008)
11. Zhou, Q., DeSantis, R.: Usability issues in city tourism web site design: a content analysis. In: IPCC 2005. Proceedings. International Professional Communication Conference, 2005., 789–796. IEEE (2005)

# Classification Models as Tools of Bankruptcy Prediction—Polish Experience

**Józef Pociecha, Barbara Pawełek, Mateusz Baryła and Sabina Augustyn**

**Abstract** The purpose of this paper is to present the results of extensive empirical research conducted in Poland on the capability of the bankruptcy prediction models most commonly used in practice to provide a correct classification. A main dataset and its variants, forming the basis for the conducted studies, were described. A comparative analysis of the most frequent bankruptcy prediction models was performed, based on the example of manufacturing companies in Poland. Empirical and theoretical distributions of financial ratios which are diagnostic variables in the presented models were examined. In summary, the conclusions were formulated allowing us to provide an answer to four basic questions raised in the introduction to the paper.

**Keywords** Bankruptcy prediction · Classification models · Ability of proper classification · Sampling methods · Distributions of financial ratios

## 1 Introduction

The bankruptcy of companies is an integral element of the market economy. However, every bankruptcy event involves both individual and social costs. For this reason, the knowledge of methods for bankruptcy prediction is sought after, and the improvement of its tools is important in both theoretical and practical terms. Basic tools of bankruptcy prediction are classification models (see e.g., [4]). Many papers have been devoted to issues of bankruptcy forecasting; their comprehensive overview was

J. Pociecha (✉) · B. Pawełek · M. Baryła · S. Augustyn
Department of Statistics, Cracow University of Economics,
27 Rakowicka Street, 31-510 Cracow, Poland
e-mail: jozef.pociecha@uek.krakow.pl

B. Pawełek
e-mail: barbara.pawelek@uek.krakow.pl

M. Baryła
e-mail: mateusz.baryla@uek.krakow.pl

S. Augustyn
e-mail: sabina.augustyn@uek.krakow.pl



F. Mola et al. (eds.), *Classification, (Big) Data Analysis and Statistical Learning*,
Studies in Classification, Data Analysis, and Knowledge Organization,
https://doi.org/10.1007/978-3-319-55708-3_18

presented, among others, by [3]. Among methods of bankruptcy prediction, the following are most commonly used in practice: multivariate discriminant analysis, logit models, neural networks, and classification trees. The subject of this paper is a presentation of some results of a comparative analysis of bankruptcy prediction models and the evaluation of their effectiveness.

With this in mind, the following questions were raised:

1. Is it possible to identify methods that perform better than others in terms of classification accuracy?
2. Do methods of sampling bankrupt and non-bankrupt companies (the pairing method and random sampling with replacement) affect the predictive ability of the models in question?
3. Do proportions of division of a dataset into a training set and a testing set affect the predictive power of the models?
4. Does the shape of empirical and theoretical distribution of financial ratios have an impact on the results of the modeling and predicting bankruptcy of companies?

The empirical studies recounted in this paper give some answers to the questions raised above.

The article presents the outcomes of empirical research which deals with the above-mentioned problems. It contains some results of a completed research project which was financed by the National Science Centre (grant No. N N111 540 140).

## 2 Dataset and Examination Variants

The empirical study of the elements which are presented in this paper was based on information obtained from the database of the EMIS (Emerging Markets Information Service), concerning companies operating in the manufacturing sector in Poland during the years 2005–2009. After eliminating from the initial dataset those companies for which the data were incomplete or outlying, a database was built that included 7329 business entities: 182 bankrupt (*B*) and 7147 non-bankrupt (*NB*) ones. Therefore, the set structure looked as follows: 2.5% of "bankrupt" and 97.5% of "financially sound" companies. The group of 182 "bankrupts" can be divided into two subgroups (*B*1 and *B*2), taking into account the year covered by the financial statements and the year in which bankruptcy was declared. In the first subgroup (*B*1), there are "bankrupts" for which financial data for the year preceding the bankruptcy are known. On the other hand, in the second subgroup (*B*2), there are "bankrupts", for which financial data were published 2 years before the bankruptcy. Another stratification of the three subgroups was their breakdown by successive years of the study. As a result, the following designations for the considered sets of companies were adopted:

B1&t—a set of "bankrupts" 1 year before the declaration of bankruptcy (*B*1), for which we have financial data of the year *t* ($t = 2006, \ldots, 2009$),

B2&t—a set of "bankrupts" 2 years before the declaration of bankruptcy ($B2$), for which we have financial data of the year $t$ ($t = 2005, \ldots, 2008$), and

NB&t—a set of "non-bankrupts" ($NB$), for which we have financial data of the year $t$ ($t = 2005, \ldots, 2009$).

Each of the companies belonging to these sets was characterized by 35 financial ratios ($R_{01} - R_{35}$) calculated on the basis of their financial statements. Indicators included in the study belong to the following groups of financial ratios:

- liquidity ratios—4 ratios ($R_{01} - R_{04}$),
- liability ratios—10 ratios ($R_{05} - R_{14}$),
- profitability ratios—8 ratios ($R_{15} - R_{22}$),
- productivity ratios—11 ratios ($R_{23} - R_{33}$), and
- other ratios—2 ratios ($R_{34} - R_{35}$).

Taking into account the best use of information contained in the available database, two variants of the examination were considered. The first one (variant $V_1$) was focused on predicting bankruptcy of companies 1 year in advance. For this purpose, due to a relatively small number of bankrupts in particular years, the longest possible time series was taken into account (i.e., the years 2006–2009), combining bankrupts of the years 2007–2010 (i.e., those who declared bankruptcy after 1 year in relation to the years 2006–2009) in one set. The same was done in the case of the second considered scenario (variant $V_2$), which is related to predicting bankruptcy of companies considering a 2-year forecast horizon. Here as well, the financial condition of companies, reflected in the financial ratios for the longest period, i.e., for the years 2005–2008, was taken into account. Therefore, this variant of the examination covered the companies which went bankrupt between 2007 and 2010 (i.e., those companies that went bankrupt after 2 years in relation to the years 2005–2008). It is also worth noting that within the framework of the examination variants $V_1$ and $V_2$, companies with a good financial standing were also combined into one set, as had been done in the case of bankrupt companies. In the conducted study, the sets of companies created from the initial database (under particular scenarios) were treated as populations of bankrupt and non-bankrupt companies for a given period, which were then used to construct relevant samples.

## 3 Comparison of the Prognostic Capabilities of the Models

In order to conduct a comparative analysis of the prognostic capabilities of the selected classification models, four types of models most commonly used in predicting bankruptcy of companies were used: a linear discriminant function, a logit model, a classification tree obtained with the CART method, and a neural network having the architecture of a multilayer perceptron (MLP network). Therefore, in the study there were considered both statistical models and techniques belonging to data mining methods, without favoring any of these two groups.

In this study, the choice of variables in the case of statistical models was made with the use of the "forward" (progressive) and "backward" (reverse) stepwise method, within the framework of a linear discriminant analysis and a logit analysis. In the case of classification trees, the selection of variables for the model was made automatically as a result of using the CART algorithm. Variables selected with the above-mentioned methods were also used to construct a neural network of MLP type.

A comparative analysis of the four types of models was conducted from the point of view of their prognostic capabilities. The basic measure of the prognostic capabilities of a model is the efficiency of the first kind, i.e., the percentage of bankrupts who have been correctly classified by the model to the set of bankrupts usually referred to as sensitivity. If models were obtained for which the efficiency of the first kind was at a similar level, then the one for which the highest value of the second kind efficiency, i.e., the percentage of non-bankrupts correctly classified by the model to the group of companies that continue their activities, that is, specificity of the model, was adopted as the best model. For comparison, the overall efficiency, which is a percentage of companies correctly classified by the model, defined alternatively as accuracy, was also shown.

The prognostic capabilities of a model were evaluated on a testing set. Consequently, the problem remains as regards the decision on division of respective sets of bankrupts and non-bankrupts into training and testing sets. In this analysis, there were included two divisions of samples into training and testing sets by a ratio of 6:4 and 7:3, that is, a variant similar to the balanced division and a variant taking into account preferences for the training set.

As indicated in the previous section, the populations of bankrupts (*B*) and non-bankrupts (*NB*) are highly unbalanced, which is natural, because in a correctly developing economy bankruptcy is rather a rare event. Therefore, there is a problem of selecting balanced samples from a highly imbalanced population of companies. A deliberate selection is the most commonly used technique, which consists in matching companies in pairs (bankrupt and non-bankrupt), called the pairing method. In this case, usually all bankrupt companies are adopted as a sample, and then companies of similar size and type of business that are in a good financial condition are looked for. An alternative to the paring method is a random sampling method. Due to a small size of the population of bankrupts, it will be sampling with replacement. The presented empirical studies were based on balanced samples, obtained with both methods, which is typical in the literature (see for example [1]). The results of the conducted research are presented in Tables 1 and 2, showing the rankings of models with the best prognostic capabilities as regards corporate bankruptcy.[1] In these tables, the two best models are presented among: discriminant models (*D*), logit models (*L*), neural networks (*NN*), and classification trees (*CT*). The numbers of the financial ratios which were included as diagnostic variables in the distinguished models are given in brackets.

[1]A part of the results presented here, concerning the application of the paring method and the random sampling with replacement method, were presented in a different paper by [2].

**Table 1** Ranking of the best models obtained for the examination variants $V_1$ and $V_2$ with companies selected by the pairing method

| Variant | Ranking | Model type | Division type | Testing set | | |
|---|---|---|---|---|---|---|
| | | | | Sensitivity | Specificity | Accuracy |
| $V_1$ | 1 | $NN(R_{14}, R_{16}, R_{31})$ | 6:4 | 95.83 | 83.33 | 89.58 |
| | 2 | $D(R_{14}, R_{16}, R_{31})$ | 6:4 | 95.83 | 75.00 | 85.42 |
| | 3 | $CT(R_{20})$ | 6:4 | 95.83 | 70.83 | 83.33 |
| | 4 | $NN(R_{05}, R_{13}, R_{24})$ | 7:3 | 94.44 | 72.22 | 83.33 |
| | 5 | $L(R_{14}, R_{16})$ | 6:4 | 91.67 | 75.00 | 83.33 |
| | 6 | $L(R_{11}, R_{31})$ | 7:3 | 88.89 | 83.33 | 86.11 |
| | 7 | $D(R_{11}, R_{14}, R_{31})$ | 7:3 | 88.89 | 77.78 | 83.33 |
| | 8 | $CT(R_{11})$ | 7:3 | 88.89 | 66.67 | 77.78 |
| $V_2$ | 1 | $NN(R_{05}, R_{11})$ | 6:4 | 83.67 | 69.39 | 76.53 |
| | 2 | $CT(R_{11})$ | 6:4 | 83.67 | 63.27 | 73.47 |
| | 3 | $NN(R_{02}, R_{11})$ | 7:3 | 81.08 | 81.08 | 81.08 |
| | 4 | $CT(R_{11})$ | 7:3 | 81.08 | 75.68 | 78.38 |
| | 5 | $L(R_{11}, R_{12})$ | 6:4 | 77.55 | 69.39 | 73.47 |
| | 6 | $L(R_{02}, R_{11})$ | 7:3 | 72.97 | 72.97 | 72.97 |
| | 7 | $D(R_{12}, R_{16}, R_{25})$ | 7:3 | 70.27 | 54.05 | 62.16 |
| | 8 | $D(R_{05}, R_{13})$ | 6:4 | 67.35 | 71.43 | 69.39 |

Conclusions from the ranking of bankruptcy prediction models presented above will be formulated in the summary included at the end of this paper.

## 4 Financial Ratios Most Often Used in the Models and Their Distributions

Let us now consider which of the explanatory variables being financial ratios occurred most often in the 32 models of bankruptcy prediction presented in the preceding paragraph. It was found that only 4 of the 35 considered ratios occurred more frequently than the others as shown in Table 3.

Next, let us ask about the empirical distributions of these ratios. Boxplots of the empirical distributions of the listed ratios for companies of the manufacturing sector in Poland are shown in Figs. 1, 2, 3, and 4. Here, attention should be paid to where the zero point of the scale is (zero value of the ratio $R_i$). In rows (on the pseudo vertical axis), the designations mentioned in Sect. 2 for the considered datasets were adopted.

At first, we can see that the empirical distributions of particular financial ratios in the bankrupt groups are more concentrated than in the non-bankrupt groups. We can see mainly right-side asymmetry of the empirical distributions in the groups of

**Table 2** Ranking of the best models obtained for the examination variants $V_1$ and $V_2$ with companies selected by random sampling with replacement

| Variant | Ranking | Model type | Division type | Testing set | | |
|---|---|---|---|---|---|---|
| | | | | Sensitivity | Specificity | Accuracy |
| $V_1$ | 1 | $NN(R_{03}, R_{16})$ | 6:4 | 100.00 | 91.67 | 95.83 |
| | 2 | $NN(R_{16}, R_{21}, R_{23}, R_{31})$ | 7:3 | 100.00 | 83.33 | 91.67 |
| | 3 | $CT(R_{20})$ | 6:4 | 100.00 | 79.17 | 89.58 |
| | 4 | $D(R_{16}, R_{31})$ | 7:3 | 100.00 | 72.22 | 86.11 |
| | 5 | $CT(R_{11})$ | 7:3 | 94.44 | 88.89 | 91.67 |
| | 6 | $L(R_{16}, R_{31})$ | 7:3 | 94.44 | 83.33 | 88.89 |
| | 7 | $D(R_{17}, R_{18}, R_{20}, R_{30})$ | 6:4 | 91.67 | 100.00 | 95.83 |
| | 8 | $L(R_{03}, R_{16})$ | 6:4 | 87.50 | 66.67 | 77.08 |
| $V_2$ | 1 | $CT(R_{11})$ | 6:4 | 89.80 | 61.22 | 75.51 |
| | 2 | $NN(R_{11}, R_{13})$ | 7:3 | 89.19 | 64.86 | 77.03 |
| | 3 | $NN(R_{02}, R_{11}, R_{13}, R_{16})$ | 6:4 | 87.76 | 67.35 | 77.55 |
| | 4 | $CT(R_{11})$ | 7:3 | 83.78 | 67.57 | 75.68 |
| | 5 | $L(R_{02}, R_{11}, R_{13}, R_{16})$ | 6:4 | 81.63 | 61.22 | 71.43 |
| | 6 | $D(R_{11}, R_{27})$ | 7:3 | 81.08 | 62.16 | 71.62 |
| | 7 | $L(R_{11}, R_{13})$ | 7:3 | 75.68 | 72.97 | 74.32 |
| | 8 | $D(R_{06}, R_{12}, R_{21}, R_{26})$ | 6:4 | 73.47 | 75.51 | 74.49 |

**Table 3** Occurrence of particular ratios in 32 best models of bankruptcy prediction

| Ratio | Number of occurrences | Definition | Ratio type |
|---|---|---|---|
| $R_{11}$ | 17 | $\frac{\text{Net profit (loss)} + \text{Depreciation}}{\text{Long-term liabilities} + \text{Short-term liabilities}}$ | liability |
| $R_{16}$ | 10 | $\frac{\text{EBITDA}}{\text{Total Assets}}$ | profitability |
| $R_{31}$ | 7 | $\frac{\text{Operating costs}}{\text{Short-term liabilities}}$ | productivity |
| $R_{13}$ | 6 | $\frac{\text{Gross profit (loss)}}{\text{Short-term liabilities}}$ | liability |

non-bankrupt firms, but also left-side asymmetry in some distributions is observed. In conclusion, we have to state that the empirical distributions of financial ratios are differentiated and strongly irregular (see histograms in Figs. 5 and 6).

In the next step, we tried to fit some theoretical distributions to empirical distributions of the main financial ratios. We tested normal, log-normal, exponential, and gamma distributions using chi-square test. The results of the testing were generally negative, except $R_{11}$ ratio distribution in group B1&2008, where the distribution is

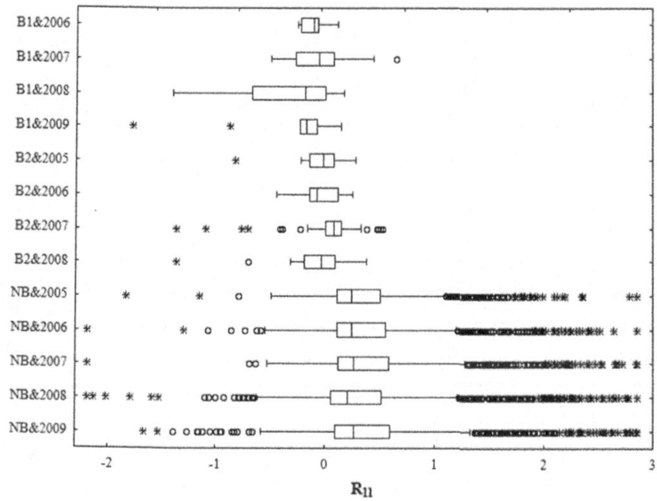


**Fig. 1** Empirical distribution of ratio $R_{11}$

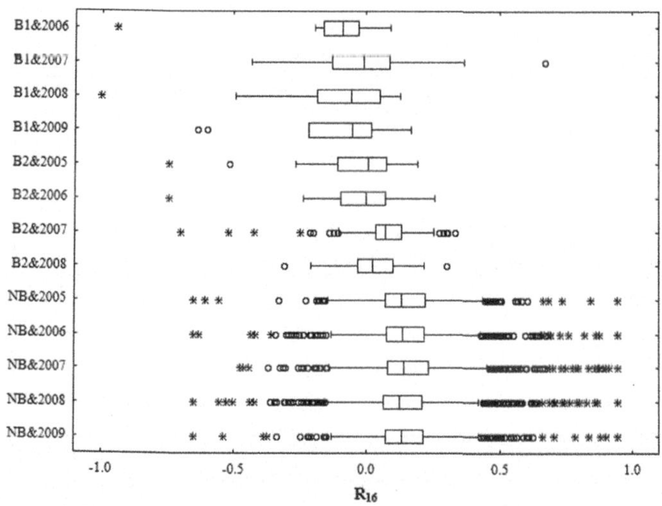


**Fig. 2** Empirical distribution of ratio $R_{16}$

normal (chi-square = 2.41, with $p$ value = 0.12) and $R_{31}$ in group B2&2007 where log-normal distribution is observed (chi-square = 5.04, $p$ value = 0.08).

The ratios most commonly used in bankruptcy predicting models have irregular distributions, often with a high asymmetry and outliers. These distributions are generally not consistent with typical theoretical distributions of random variables.

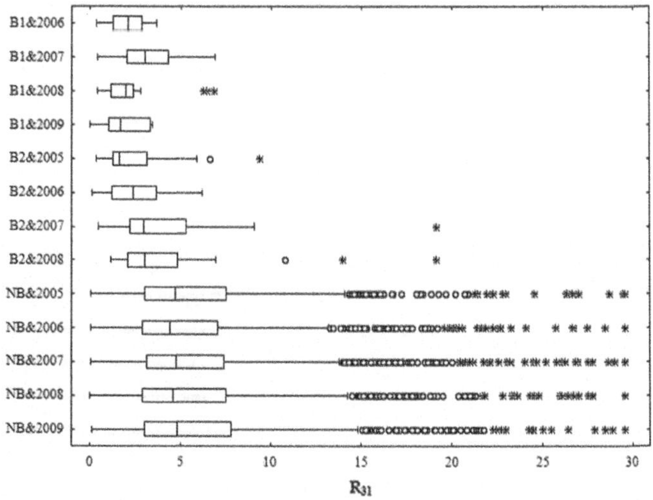


**Fig. 3** Empirical distribution of ratio $R_{31}$

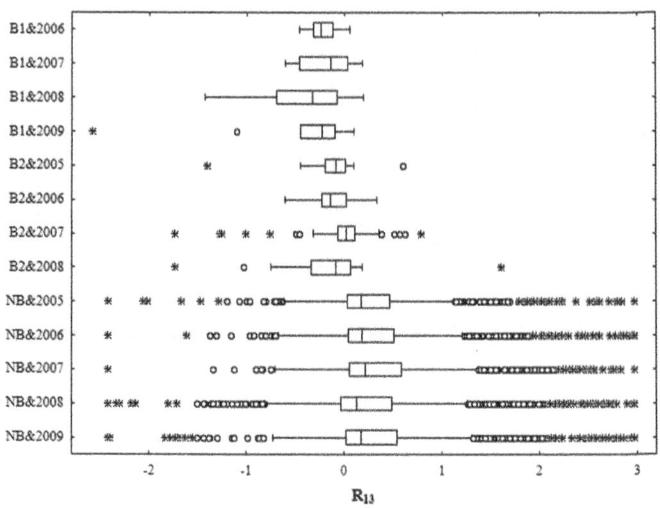


**Fig. 4** Empirical distribution of ratio $R_{13}$

## 5 Summary

The results of the conducted comprehensive empirical studies presented in Sects. 3 and 4 allow us to formulate the following answers to the questions raised in the introduction:

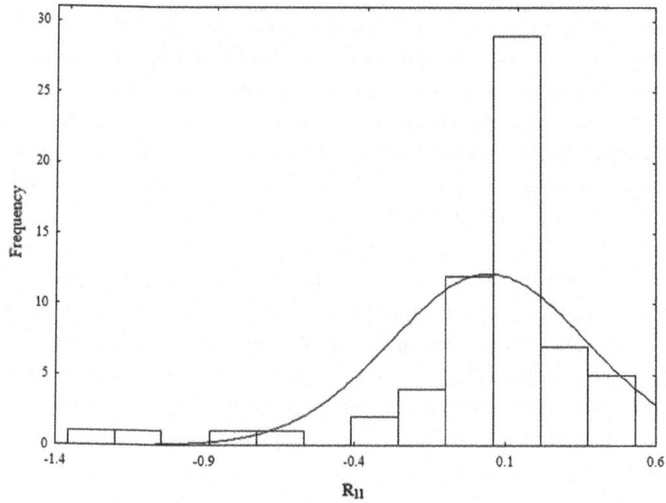


**Fig. 5** Empirical distribution of ratio $R_{11}$ in group B2&2007

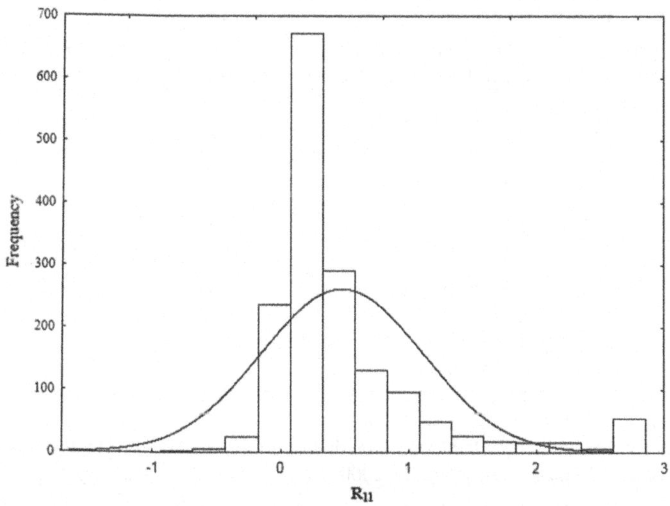


**Fig. 6** Empirical distribution of ratio $R_{11}$ in group NB&2007

(1) All in all it cannot be unambiguously concluded that a certain method of bankruptcy prediction is more useful than others. It depends on the fact that the relationships between the financial ratios are nonlinear. The results of the investigation indicate a clear advantage of machine learning methods over discriminant or logit models.

(2) The method of sampling affects the prognostic capabilities of the models. Models with a higher predictive capability were generally obtained from random samples rather than from pairing methods most often used in practice.
(3) There is no unambiguous answer to the question: what the best ratio of division of a dataset into training and testing sets. However, the results of empirical studies indicate an advantage of a more balanced division over a division giving a clear advantage to a training set.
(4) The most difficult part is to formulate an answer to the fourth question. The financial ratios being diagnostic variables in the considered classification models have irregular empirical distributions with both right-tailed and left-tailed asymmetries. As a rule, none of the more common theoretical distributions can be matched up to them. In particular, it should be noted that the distributions of financial ratios diverge from a normal distribution. This is a reason for the advantage of nonparametric classification models over classical parametric models.
To sum up, it should be stated that the conducted study on the efficiency of the bankruptcy prediction models clearly illustrates the difficulty in reconciling theoretical assumptions underlying data classification methods with the economic reality of the analyzed datasets.

**Acknowledgements** The publication was financed from the funds granted to the Faculty of Management at Cracow University of Economics, within the framework of the subsidy for the maintenance of research potential.

## References

1. Altman, E.I., Hotchkiss, E.: Corporate Financial Distress and Bankruptcy. Wiley, Inc. (2005)
2. Baryła M., Pawełek B., Pociecha J.: Selection of Balanced Structure Samples in Corporate Bankruptcy Prediction. In: Wilhelm, A.F.X., Kestler,H.A. (eds.) Analysis of Large and Complex Data, Springer, pp. 345–355 (2016)
3. Bellovary, J., Giacomino, D., Akers, M.: A review of bankruptcy prediction studies: 1930 to present. J. Financial Educ. **33**, 1–42 (2007)
4. McKee, T.E.: Developing a bankruptcy prediction model via rough sets theory. Int. J. Intell. Syst. Acc. Fin. Manag. **9**(3), 159–173 (2000)

# Quality of Classification Approaches for the Quantitative Analysis of International Conflict

**Adalbert F. X. Wilhelm**

**Abstract** We provide an evaluative comparison of some modern classification algorithms, such as CART, AdaBoost, bagging and random forests, to predict the incidences of military conflicts and other political relevant events. Our evaluative comparison is based on two main aspects: the importance of variables within the classifier as well as the prediction accuracy. While modern classification procedures are able to improve the prediction accuracy as compared to the traditionally used logistic regression, the logistic regression still holds a large advantage in terms of interpretability of the variables' relevancy.

**Keywords** Logistic regression · Classification trees · Boosting · Rare events

## 1 General Purpose

The increase in available data about military conflicts, e.g., collected by the correlates of war project [8] or the Uppsala Conflict Data Program [11], provides the political science researcher with ample opportunities to study the onset of wars, armed conflicts, and other politically relevant events. This has led to a significant increase in application papers using quantitative analysis techniques to model armed conflicts as one major aspect of international relations. From a methodological point of view, the logistic regression has become kind of a panacea for these analyses and rests at the cornerstone in this field. Robust estimation correcting for heteroscedasticity and the cluster structure of the panel data have become standard specifications over the past years. Since the major aim in political science targets the empirical evaluation of theories and hence the explanatory aspects of statistical modeling, the preference for the logistic regression approach is a logical consequence. The rare cases in which also the models' predictive aspects have been mentioned show that these regression models perform rather badly in this aspect. The fact that military conflicts constitute rare events in these data sets is fortunate from a humanitarian point of view. It poses,

A. F. X. Wilhelm (✉)
Jacobs University Bremen, Campus Ring 1, D-28759 Bremen, Germany
e-mail: a.wilhelm@jacobs-university.de

 
F. Mola et al. (eds.), *Classification, (Big) Data Analysis and Statistical Learning*, Studies in Classification, Data Analysis, and Knowledge Organization,
https://doi.org/10.1007/978-3-319-55708-3_19

however, methodological challenges in the form of class imbalance, see [4]. However, in the typical data sets analyzed for publications incidence rates of conflicts lie above the actual incidence by focussing on specific world regions and time periods and hence shift toward more balanced classes.

Over the last decade, a number of modern classification algorithms, such as CART, AdaBoost, neural nets, support vector machines, and random forests, have been proposed. Here, we compare the traditional logistic regression approach to some modern classification methods as implemented in the `caret` package [5] in `R` [9].

## 2 Data Sets

The data sets for this evaluation study are chosen in such a way that they cover some of the most prominent data collection and analysis projects in the field of international relations. A brief overview on some of the key features of the three data sets is provided in Table 1.

**Sub-Saharan Africa I**. Over the last decade, there have been numerous political science studies dealing with civil wars in Sub-Saharan Africa. As can be seen in the treemap (see Fig. 1), the distribution of civil war incidences during the 1980 and 1990s is quite skewed: out of 43 countries, 29 countries experienced civil conflict at least within 1 year; quite some countries, such as South Africa, Chad, Ethiopia, Angola, and Sudan suffered from civil–military conflicts almost throughout this period. A total of 14 states had not been haunted by civil conflict at all during these years. Civil conflicts during this period not only caused numerous lives during fighting but also lead to massive displacement from their homes for many people and caused long-lasting diseases and disabilities which impacted the lives of various generations [3] and hampered the economic development. See [10] for a review on research that addresses the association between economic conditions and civil conflict.

The first data set we have chosen covers arms trade, military expenditure, and their impact on the occurrence of armed conflicts in sub-Saharan Africa. The data set has been prepared and originally analyzed by [2] and is available in the replication data archive of the *Journal of Peace Research.*[1]

A primary goal of the original research with this data was to determine whether arms trade (`ARMSTRAN`) is a predictor of political violence in sub-Saharan Africa (cf. [2], p. 696) over and above the effect of military expenditure (`MILSPEND`). In case of a positive assertion to this first hypothesis, a second question is whether arms trade's influence is already covered by the effect of military expenditure (`MILSPEND`) or whether it adds additional information. Common theories on military conflicts agree on the tendency of instable regimes to be more likely involved in military conflicts. Instability of a regime is here operationalized by variables on regime transition (`TRANSITI`), the presence of ethno-political groups (`ETHNOPOL`), on political vio-

[1]Data can be found at http://legacy.prio.no/Research-and-Publications/Journal-of-Peace-Research/Replication-Data/Detail?oid=301968.

**Table 1** Summary information about the data sets used in the evaluative comparison

| Data set | # cases | # countries | # time periods | Incidence rate |
|---|---|---|---|---|
| Sub-Saharan Africa I | 1017 | 46 | 26 | 0.22 |
| Sub-Saharan Africa II | 743 | 41 | 19 | 0.27 |
| Petrostates | 7768 | 188 | 59 | 0.17 |

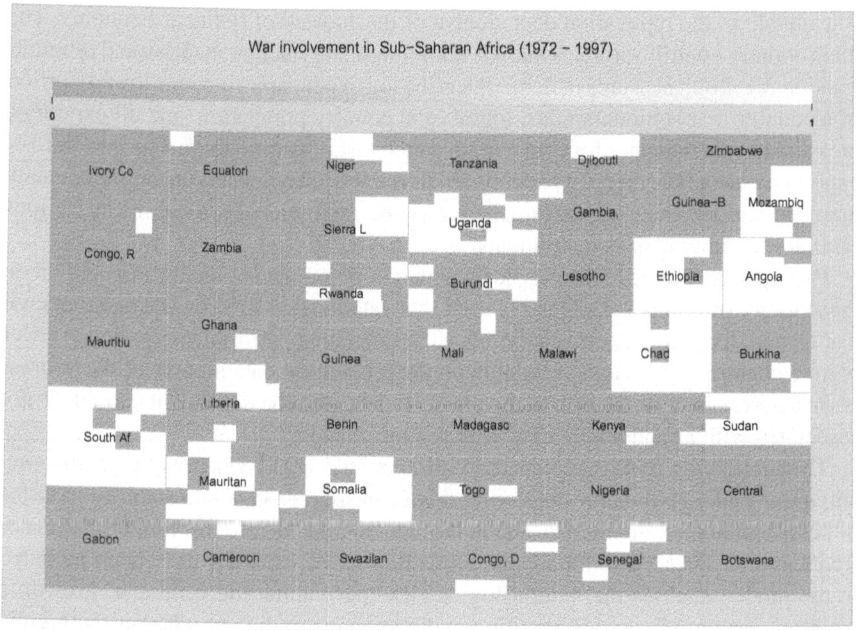


**Fig. 1** A treemap showing incidence of civil war in Sub-Saharan Africa in the period of 1972 to 1997. White cells indicate years in which a civil war was ongoing in the country. The size of each cell represents the per capita GNP of the countries in each year. Angola, Chad, Ethiopia, and South Africa suffered from civil war almost throughout all the years under investigation. Only 14 states in this region, including Mauritius, Gabon, and Ivory Coast, have not been involved in civil wars during that time

lence exerted by the government (`REPRESSI`), and semi-democracy (`SEMIDEM`). Economic development (`DEVELOP`) is widely seen as an influential factor for keeping peace. The variable `CUMWAR` has been included to control for autocorrelation of war involvement. Some authors have argued that the colonial legacy (`COLONIAL`) also plays a role in this context due to the different styles of ruling in the past and the resulting different possibilities of political organization of minorities and political opponents.

**Sub-Saharan Africa II**. The second data set deals with the same region but focuses on the relationship between economic development and armed conflicts in this region. An armed conflict is defined in the PRIO/Uppsala database as "a contested incompatibility which concerns government and/or territory where the use of armed force between two parties, of which at least one is the government of a state, results in at least 25 battle-related deaths." According to this definition, armed conflicts are a superset of civil wars. The data set has been originally analyzed by [6] and is available in the replication data archive of the *Journal of Political Economy*. The main country control variables include ethnolinguistic fractionalization and religious fractionalization, measures of democracy, the log of per capita income, the proportion of a country that is mountainous, log of total country population, and oil exporters, measured by an indicator for countries in which oil constitutes more than one-third of export revenues. The original paper by [6] puts a special emphasis on the endogeneity question and includes an instrumental variable regression approach using various rainfall measurements as instruments.

**Petrostates**. The third data set deals with non-democracies and the effect of oil on the stability of regimes. Colgan [1] used this data to shed light on the fact that civil war is more frequent in petrostates and at the same time petrostates experience fewer regime changes. The data is available on the replication data archive of the *Journal of Peace Research*. In contrast to the previous data sets, the dependent variable is not associated with armed conflict but with regime change.

The underlying political science question treated in [1] concerns two competing empirical findings: the one suggesting that oil resources generate stability in terms of persistent autocratic regimes and at the same time generate instability in terms of frequent civil wars. Additionally, a couple of control variables have been included in the model, such as the natural log of GDP per capita, the economic growth, a country's history of regime transition, a dichotomous indicator whether the regime is a monarchy, the percentage of muslim population, and a series of period dummies "to control for temporal patterns and contemporaneous shocks" ([1], p. 11).

## 3 Evaluative Comparisons

As mentioned before, the typical analysis in this application field aims at explaining the relationship between various potential predictors and the occurrence of conflict. The potential predictors are mainly associated with specific aspects of political theories and the quantitative analysis is performed to provide empirical corroboration for political theories and certain policies. Typically, prediction is not in the major focus of the analysis. The rare cases in which also the models' predictive aspects have been mentioned show that the logistic regression models perform rather poorly in this aspect. King and Zeng [4] discuss methodological consequences in this field for rare events. The data sets we are looking at, however, show incidence rates that appear large enough to avoid any problematic issues here.

### 3.1 Evaluation Design

Our evaluative comparison is based on two main aspects: the importance of variables within the classifier and the prediction accuracy. We use the area under the curve as our main measure of predictive accuracy. In addition, the ROC curves as well as the specificity of the classifiers will be used to differentiate between classifiers in more details. We split our data into training and test data in such a way that roughly the first 75% of data points have been used in the training data while the latter 25% have been assigned to the test data. For the Sub-Sahara I data set, this means that country-years up to the year 1991 have been used as training cases. For the data Sub-Saharan Africa II, the split was performed in 1994, while for the Petrostate data it was done in 1990. This split was performed to keep the temporal dependency and to prevent the classifiers to make use of additional structure obtainable from future cases.

For each method, we used ten-fold cross-validation on the training data to obtain the final model. While logistic regression produces a single model by default, the other three methods search over various tuning parameters for the best model. The tuning parameters that have been used are the number of trees and the maximum tree depth for bagged Adaboost (`adabag`). The boosting method `Adaboost.M1` uses the same two tuning parameters as `adabag` and in addition different versions of the weight updating coefficient. The chosen `caret` implementation of random forests optimizes over the number of randomly selected predictors.

### 3.2 Evaluation Results

We use the Receiver Operating Characteristic (ROC) curves as basis for comparing the models. As a single metric, the area under the curve (AUC) is widely used to evaluate the predictive performance of classifiers. Table 2 summarizes the AUC for the various instances. The ROCs are given in Fig. 2. As we can see, the boosting method consistently outperforms logistic regression and the bagging model. Random forests obtain the best results for the first two data sets, especially for the data set Sub-Saharan II the resulting AUC is by far larger than for the other three methods.

**Table 2** Predictive accuracy for some classification techniques as measured by AUC (area under the ROC curve) on presented data sets

| Data set | Logistic | AdaBag | AdaBoost | Random Forests |
|---|---|---|---|---|
| Sub-Saharan Africa I | 0.8307 | 0.8908 | 0.8481 | 0.8605 |
| Sub-Saharan Africa II | 0.6017 | 0.743 | 0.7551 | 0.8078 |
| Petrostates | 0.7086 | 0.5 | 0.7124 | 0.6048 |

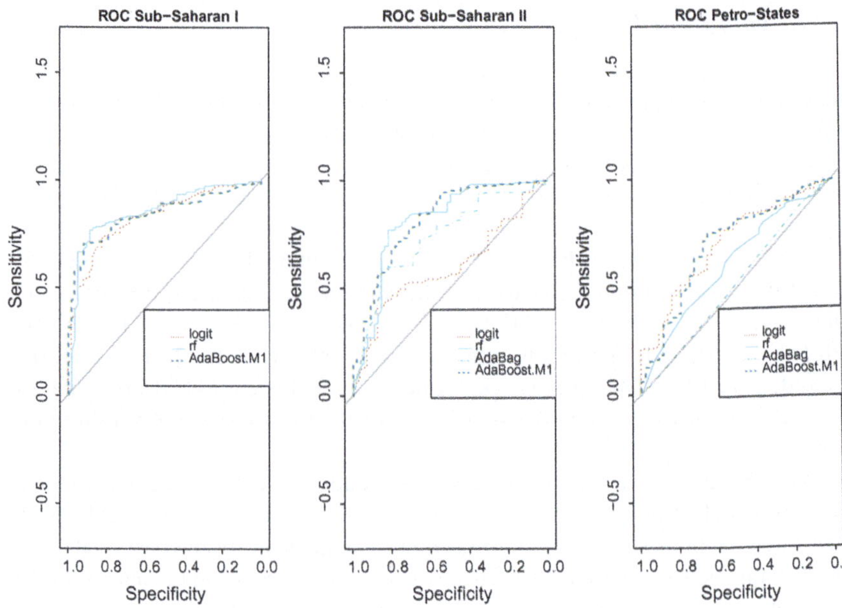


**Fig. 2** ROC curves for the four classification methods evaluated for the three data sets. In the right plot showing the results for the Petrostate data, the rather poor performance of random forests and bagged trees is clearly visible

However, for the Petrostate data, random forests perform even worse than logistic regression.

In the given situations, predictions of the occurrence of war, armed conflict, or regime change are the major purposes. Hence, accuracy in correctly predicting these events is typically of larger importance than correctly predicting the peace cases. When looking at the sensitivity measures (cf. Table 3), one can see that the boosting method and random forests perform on a similar level and clearly outperform the other two methods in this respect. The Petrostate data example, which shows the lowest incidence rate among the three case studies, is however not well predicted by any of the methods. Here, a closer investigation of the class imbalance setting is needed. The current model specifications have not been adjusted to that as the incidence rates in the data sets used have been around 20%, see Table 1.

The second major goal of the evaluation was to check whether machine learning algorithms would also identify the same predictor variables as the logistic regression models. The well-established assessment of significant predictors in logistic regression via hypothesis testing lacks a direct analogue for machine learning algorithm, but the resampling procedures allow to use variable importance measures as substitute. Evaluation of variable importance is done by comparing the results of the generic function `varImp` from the `R` package `caret`. This function uses different

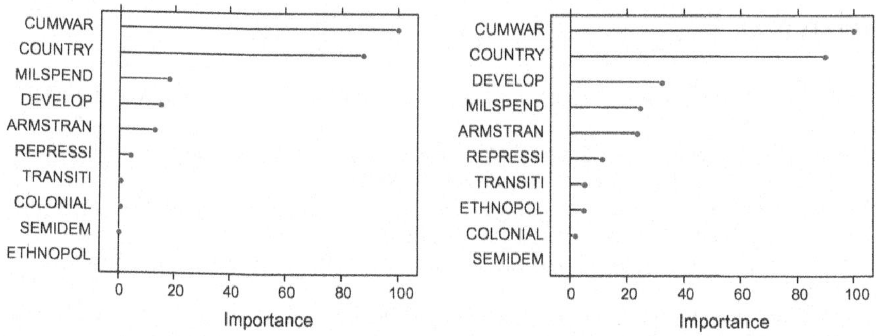


**Fig. 3** Variable importance plot to compare the explanatory aspects of the various classifiers. The plots above relate to the data set `Sub-Saharan Africa I`; the plot on the left shows the results for the AdaBoost models, the one on the right for the random forest models

**Table 3** Sensitivity (recall or true positive rate) measuring the percentage of civil wars, armed conflicts or regime changes, respectively, in the three presented case studies that have been correctly predicted in the test data by the four classifiers examined

| Data set | Logistic | AdaBag | AdaBoost | Random Forests |
|---|---|---|---|---|
| Sub-Saharan Africa I | 0.7713 | 0.7717 | 0.8151 | 0.8026 |
| Sub-Saharan Africa II | 0.1743 | 0.6171 | 0.8181 | 0.8043 |
| Petrostates | 0 | 0 | 0.0196 | 0.0067 |

measures for each model. For details, please see [5]. The results can be summarized graphically, see Fig. 3.

The variable importance measures extract the same predictor variables as the corresponding logistic regression models and provide a reasonable alternative to significance testing.

## 4 Conclusion

As we can conclude from the above evaluations boosting models and random forests achieve better results than the logistic regression model with respect to predictive accuracy for the class of interest. The results differ in the three case studies examined and require a detailed analysis of the class imbalances. In contrast to widespread reservations, the machine learning approaches are well able to identify the important predictor variables which then can be further investigated for causal mechanisms and explanatory interpretation, see also [7]. Additional investigations are needed to incorporate the panel structure of the data into the machine learning approaches by using random effects models.

## References


1. Colgan, J.D.: Oil, domestic conflict, and opportunities for democratization. J. Peace Res. **52**(1), 3–16 (2015)
2. Craft, C., Smaldone, J.P.: The arms trade and the incidence of political violence in sub-Saharan Africa, 1967–97. J. Peace Res. **39**(6), 693–710 (2002)
3. Ghobarah, H.A., Huth, P., Russett, B.: Civil wars kill and maim people-long after the shooting stops. Am. Political Sci. Rev. **97**(2), 189–202 (2003)
4. King, G., Zeng, L.: Explaining rare events in international relations. Int. Organ. **55**(6), 693–715 (2001)
5. Kuhn, M.: Building predictive models in R using the caret package. J. Stat. Softw. **28**(5), 1–26 (2008)
6. Miguel, E., Satyanath, S., Sergenti, E.: Economic shocks and civil conflict: an instrumental variables approach. J. Political Econ. **112**(4), 725–753 (2004)
7. Muchlinski, D., Siroky, D., He, J., Kocher, M.: Comparing random forest with logistic regression for predicting class-imbalanced civil war onset data. Political Anal. **24**(1), 87–103 (2016)
8. Palmer, G., D'Orazio, V., Kenwick, M., Lane, M.: The mid4 data set: Procedures, coding rules, and description. Confl. Manag. Peace Sci. **32**(2), 222–242 (2015)
9. R Core Team: R—A Language and Environment for Statistical Computing. R Foundation for Statistical Computing, Vienna, Austria (2014)
10. Sambanis, N.: A review of recent advances and future directions in the quantitative literature on civil war. Def. Peace Econ. **13**(3), 215–243 (2002)
11. Themnér, L., Wallensteen, P.: Armed conflicts, 1946–2013. J. Peace Res. **51**(4), 541–554 (2014)

# Part VI
# Time Series and Spatial Data

# P-Splines Based Clustering as a General Framework: Some Applications Using Different Clustering Algorithms

**Carmela Iorio, Gianluca Frasso, Antonio D'Ambrosio and Roberta Siciliano**

**Abstract** A parsimonious clustering method suitable for time course data applications has been recently introduced. The idea behind this proposal is quite simple but efficient. Each series is first summarized by lower dimensional vectors of P-spline coefficients and then, the P-spline coefficients are partitioned by means of a suitable clustering algorithm. In this paper, we investigate the performance of this proposal through several applications showing examples within both hierarchical and non-hierarchical clustering algorithms.

**Keywords** P-spline · Clustering · Unsupervised learning · Time course data

## 1 Introduction

Time course data (i.e., time series and gene expression data) arise in many scientific areas and in the last years several time series clustering techniques have been proposed. For a detailed and complete literature review on this topic, we refer to [1].

For example, [3] and [2] introduced, respectively, a hierarchical clustering framework for the analysis of stationary time series and a scale-invariant distance function designed for Gaussian distributed errors. In [4], the use of k-means algorithms for

C. Iorio (✉) · R. Siciliano
Department of Industrial Engineering,
University of Naples Federico II, Napoli, Italy
e-mail: carmela.iorio@unina.it

R. Siciliano
e-mail: roberta@unina.it

G. Frasso
Faculté des Sciences Sociales, University of Liége, Liége, Belgium
e-mail: gianfrasso@gmail.com

A. D'Ambrosio
Department of Economics and Statistics,
University of Naples Federico II, Napoli, Italy
e-mail: antdambr@unina.it



F. Mola et al. (eds.), *Classification, (Big) Data Analysis and Statistical Learning*,
Studies in Classification, Data Analysis, and Knowledge Organization,
https://doi.org/10.1007/978-3-319-55708-3_20

functional data was investigated. Baragona [5] proposed different meta-heuristic methods to cluster a set of time series. In order to reduce the dimensionality of the time series clustering problem, [6] introduced the perceptually important point (PIP) algorithm. Both k-means and k-medoid algorithms for misaligned functional curves were proposed by [7]. The clustering by dynamics (CBD) Bayesian framework has been presented by [8].

When dealing with time series data, the dimensionality of the clustering task becomes easily challenging and a dimensionality reduction pre-processing step is required. Furthermore, the available measurements can be observed over different time domains and the treatment of missing observations becomes often crucial. In order to efficiently overcome these issues, [9] presented a parsimonious time series clustering framework based on P-splines [10]. The idea behind this proposal is quite simple but efficient: they model each series by P-spline smoothers and perform a cluster analysis on the estimated coefficients. P-splines allow to summarize the observed series by lower dimensional vector of parameters representing the "skeleton" of the final fit [11]. This property is not shared by different smoothers or by P-splines built on different basis functions (see e.g., [12, 13]). In this paper, we evaluate the performances achieved by this approach within both hierarchical and nonhierarchical clustering algorithms.

## 2 P-Spline Based Clustering in a Nutshell

P-splines have been introduced by [10] as flexible smoothing procedures combining B-spline bases (see e.g., [14]) and difference penalties. Let $\{x, y\}_{j=1}^{n_i}$ be a set of data, where the vector $\boldsymbol{x}$ indicates the independent variable (e.g., time) and $\boldsymbol{y}$ the dependent one. The aim is to describe the available measurements through an appropriate smooth function. We assume that the observed series can be modeled as

$$\mathbf{y} = \mathbf{f}(\mathbf{x}) + \boldsymbol{\epsilon}, \tag{1}$$

where $\boldsymbol{\epsilon}$ is a vector of errors and $f(\cdot)$ is an unknown smooth function. Denote with $\boldsymbol{B}_h(x; q)$ the value of the $h$th B-spline of degree $q$ defined over a domain spanned by $m$ equidistant knots. A curve that fits the data is given by $\hat{y}(x) = \sum_h a_h \boldsymbol{B}_h(x; q)$ where $a_h$ (with $h = 1, ..., m + q$) are the B-splines coefficients estimated through least squares. Because of over-fitting problems, $\boldsymbol{a}$ can be estimated in a penalized regression setting:

$$\hat{\boldsymbol{a}} = \underset{a}{\operatorname{argmin}} \|\boldsymbol{y} - \boldsymbol{B}\boldsymbol{a}\|^2 + \lambda \|\boldsymbol{D}_d \boldsymbol{a}\|^2, \tag{2}$$

where $\lambda$ is the positive parameter controlling the degree of smoothness of the final fit and $\boldsymbol{D}_d$ is a $d$th-order difference penalty matrix such that $\boldsymbol{D}_d\boldsymbol{a} = \Delta^d \boldsymbol{a}$ defines a vector of $d$th-order differences of the coefficients. Popular methods for smoothing parameter selection are the Akaike information criterion, cross validation, and generalized cross validation [15]. In the presence of correlated errors, the V-curve criterion introduced by [16] is a useful alternative. The P-spline based clustering [9] suggests to perform a cluster analysis on the estimated lower dimensional P-spline coefficients (as they represent the skeleton of the final fit). Furthermore, the properties of this smoother allow for an efficient treatment of missing observations [11]. Iorio [9] carried out different simulation studies to investigate the influence of the optimal number of basis as well the optimal value of smoothing parameter on the estimated P-splines showing as the final estimates ensures an efficient interpolation of the observed data. For a detailed and rigorous discussion of procedure of the P-spline based clustering procedure, one can refer [9].

## 3 Some Experiments on Real Data Sets

In this section, we show some applications of the P-spline based clustering approach on different data sets. Each series in the data sets has been modeled by P-splines taking cubic bases and third-order penalties. The amount of the shrinkage was always selected through the V-curve procedure [16]. This criterion does not require the computation of the effective model dimension which can become time consuming for long data series and offers a valuable simplification of the searching criterion by requiring the minimization of the Euclidean distance between the adjacent points on the L-curve proposed by [17]. Formally, given a P-spline defined for a fixed $\lambda$, we compute

$$\{\omega(\lambda); \theta(\lambda)\} = \left\{ \left\| y - B\hat{a}(\lambda) \right\|^2 ; \left\| D\hat{a}(\lambda) \right\|^2 \right\}$$

$$\{\psi(\lambda); \phi(\lambda)\} = \{log(\omega); log(\theta)\}$$

The L-curve is a plot of $\psi(\lambda)$ versus $\phi(\lambda)$ parameterized by $\lambda$. This plot shows a corner in a region characterized by intermediate values of $\psi$, $\phi$, and $\lambda$. The optimal amount of smoothing corresponds to the corner, namely the point of maximum curvature computed by first and second derivatives. The V-curve criterion selects the optimal smoothing with the point satisfying

$$\min \sqrt{\Delta(\phi(\lambda))^2 + \Delta(\psi(\lambda))^2}, \tag{3}$$

where $\Delta$ indicates the first-order difference operator. Thus, the selection procedure of $\lambda$ was simplified.

The optimal number of clusters has been always chosen using the GAP statistics proposed by [18].

## 3.1 Hierarchical Clustering

As a first example, we analyzed the *Drosophila melanogaster gene expression data* discussed in [19]. Three genes categories can be distinguished according to their expression profiles. The data set counts 23 muscle-specific, 33 eye-specific, and 21 transient early zygotic genes measured over 58 sequential time points (from fertilization to aging adults) with sparse missing observations. Each gene expression profile has been smoothed using cubic B-splines defined over 25 equally spaced interior knots. A hierarchical clustering algorithm with the average linkage criterion was performed in combination with a Penrose shape distance [20] in order to partition the estimated P-spline coefficients. Figure 1 shows the results achieved by our clustering procedure. As the composition of clusters is known, we used the Adjusted Rand Index (ARI) as external clustering validation criterion [21]. With this setting, the ARI was found equal to 0.9595, showing results very similar to the ones achieved by [9] and [22] which use partitioning algorithms.
Hierarchical clustering was also performed on the *phoneme* data set (R-package `fda.usc`, [23]). The data have been acquired by selecting five phonemes based on digitized speech frame and consist of 250 series of length 150 with five known classes (phoneme). Each speech frame has been smoothed by P-spline defined over 50 equally spaced interior knots. We compared the performances of the P-spline based clustering procedure with those shown in [23] by using a subset of *phoneme* data (three classes and 150 series in analogy with what presented in `fda.usc`). Figure 2 shows the dendrogram obtained by using the average linkage criterion and the Euclidean distance. The ARI was found equal to 0.9799, which is similar (even though slightly larger) to those obtained by reproducing the example in [23].

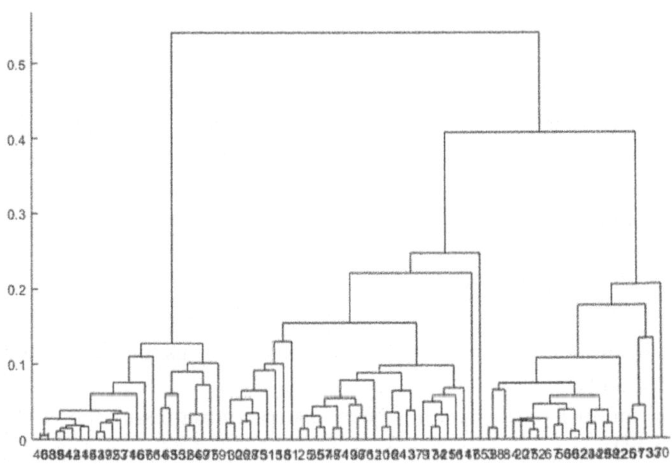


**Fig. 1** Hierarchical clustering result for *drosophila melanogaster life cycle gene expression* data

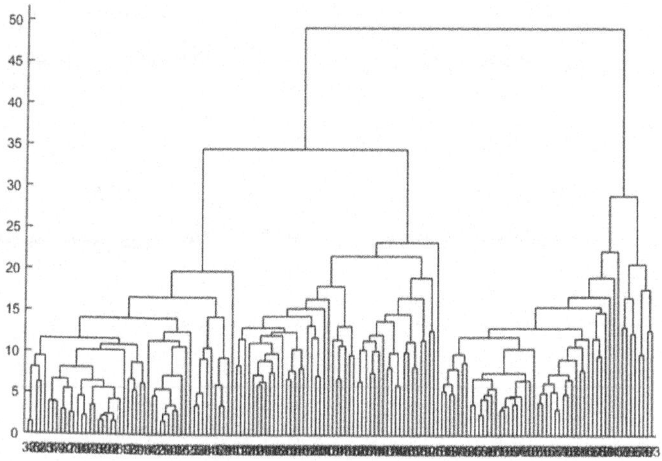


**Fig. 2** Hierarchical clustering result for *phoneme* data

## 3.2 *Partitioning Algorithms*

As a first example of applications of the P-spline based clustering procedure with non-hierarchical clustering methods, we analyze the complete *phoneme* data set (five disjoint phoneme classes) described above. The P-spline smoothers were set as in the previous case. We performed a k-medoid algorithm [24] by using the Penrose shape distance. As internal validation criterion, we used the GAP statistics [18] with 200 bootstrap replications and center numbers from 1 to 6. Table 1 shows the values of both GAP and standard errors indicating the optimal number of cluster equal to 5 according to the 1-SE rule. The ARI of the k-medoid algorithm is equal to 0.7127, indicating a good degree of concordance between theoretical and estimated clusters. Figure 3 shows the true clusters (gray curves) and the estimated cluster centers as summarized by the P-spline coefficients (black dots). It is remarkable that the shape of the series assigned to each cluster is sufficiently homogeneous even if the recognized clusters are not perfectly equal to the ones known a priori. By analyzing the same data, the functional k-means algorithm proposed by [23] achieved a lower performance (ARI = 0.2571). Finally, hierarchical clustering

**Table 1** GAP statistics computed for k-medoid partitioning applied to the phoneme data (200 bootstrap samples)

| Number of centers | 1 | 2 | 3 | 4 | 5 | 6 |
|---|---|---|---|---|---|---|
| GAP | 0.4229 | 0.6640 | 0.9735 | 1.0484 | 1.1338 | 1.1534 |
| SE | 0.0296 | 0.0287 | 0.0250 | 0.0285 | 0.0283 | 0.0277 |

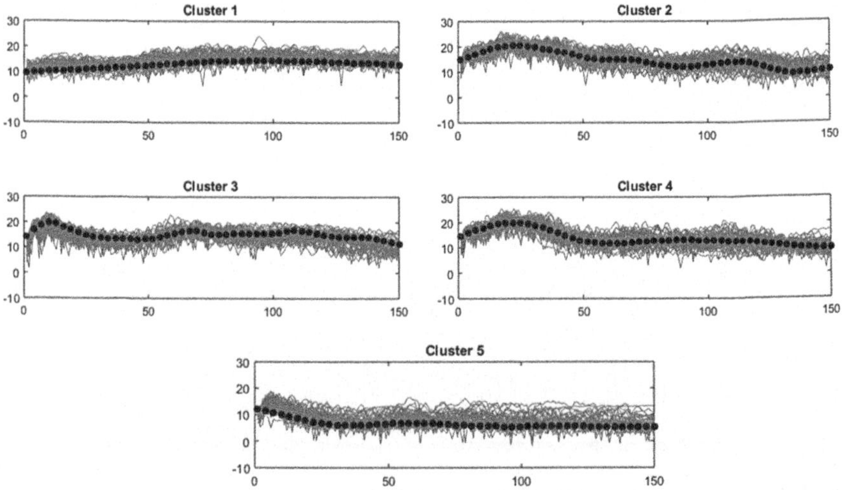


**Fig. 3** K-medoid clustering result for *phoneme* data. Dashed lines are the observed series as well as black dots represent medoids spline coefficients

**Table 2** GAP statistics computed for k-medoid partitioning applied to the Synthetic Chart Control data (200 bootstrap samples)

| Number of centers | 1 | 2 | 3 | 4 | 5 | 6 | 7 | 8 |
|---|---|---|---|---|---|---|---|---|
| GAP | 0.5570 | 0.8181 | 1.3801 | 1.4334 | 1.5378 | 1.6368 | 1.6479 | 1.6502 |
| SE | 0.0241 | 0.0159 | 0.0222 | 0.0181 | 0.0173 | 0.0204 | 0.0201 | 0.0190 |

procedures did not give satisfactory results on the complete phoneme data set (not shown).

A non-hierarchical cluster analysis was also performed for the *synthetic control chart time series data set* (freely available from the UCI Data Archive, [25]). The data matrix contains 600 synthetically generated control charts [26]. The observations can be grouped in six balanced classes: normal, cyclic, increasing, decreasing, upward, and downward. We performed a k-medoid analysis based on the standardized Euclidean distance. The (1-SE based) GAP statistics computed over 200 bootstrap replications suggested $k = 6$ as optimal number of cluster centers (see Table 2). The raw data have been modeled by P-splines built over 30 internal knots. The ARI of the obtained partition is equal to 0.7362, showing a good degree of concordance. Figure 4 shows the true observed clusters (gray curves) and the estimated cluster centers (spline coefficients, black dots). In our opinion, these results are particularly encouraging since these particular data are usually analyzed within classification rather than clustering tasks [27].

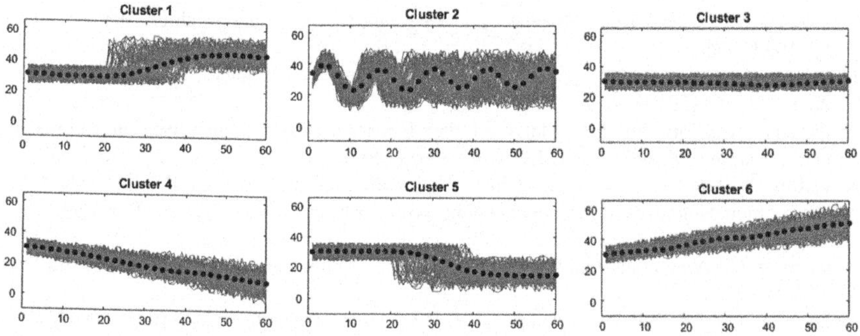


**Fig. 4** K-medoid clustering result for *Synthetic Chart Control* data. Dashed lines are the observed series as well as black dots represent medoids spline coefficients

# 4 Conclusion

In this paper, we have shown the performances of the P-spline based approach proposed by [9] by dealing with different clustering procedures. In particular, the merits of this proposal have been evaluated within both non-hierarchical and hierarchical settings. The comparisons with alternative approaches have highlighted the appropriateness of the discussed method in several real data analyzes. On the other hand, the choice of the partitioning metric still remains a crucial issue in cluster analysis applications. Nevertheless, the P-spline based clustering procedure seems to be a flexible tool to be used in alternative (or jointly) to other well-known techniques.

# References

1. Liao, T.W.: Clustering of time series data: a survey. Pattern Recogn. **38**(11), 1857–1874 (2005)
2. Maharaj, E.A.: Cluster of time series. J. Classif. **17**(2), 297–314 (2000)
3. Kumar, M., Patel, N.R., Woo, J.: Clustering seasonality patterns in the presence of errors. In: Proceedings of the Eighth ACM SIGKDD International Conference on Knowledge Discovery and Data Mining, ACM, 557–563 (2002)
4. García-López, M.L., García-Ródenas, R., Gómez-González, A.: K-means algorithms for functional data. Neurocomputing **151**, 231–245 (2015)
5. Baragona, R.: A simulation study on clustering time series with metaheuristic methods. Quad. di Stat. **3**, 1–26 (2001)
6. Fu, T.C., Chung, F.L., Ng, V., Luk, R.: Pattern discovery from stock time series using self-organizing maps. In: Workshop Notes of KDD2001 Workshop on Temporal Data Mining, ACM SIGKDD, 26–29 (2001)
7. Sangalli, L.M., Secchi, P., Vantini, S., Vitelli, V.: Functional clustering and alignment methods with applications. Commun. Appl. Ind. Math. **1**(1), 205–224 (2010)
8. Ramoni, M., Sebastiani, P., Cohen, P.: Bayesian clustering by dynamics. Mach. Learn. **47**(1), 91–121 (2002)
9. Iorio, C., Frasso, G., D'Ambrosio, A., Siciliano, R.: Parsimonious time series clustering using P-splines. Exp. Syst. Appl. **52**, 26–38 (2016)

10. Eilers, P.H.C., Marx, B.D.: Flexible smoothing with B-splines and penalties. Stat. Sci. **11**(2), 89–102 (1996)
11. Eilers, P.H.C., Marx, B.D.: Splines, knots, and penalties. Wiley Interdisc. Rev. Comput. Stat. **2**(6), 637–653 (2010)
12. Abraham, C., Cornillon, P.A., Matzner-Løber, E., Molinari, N.: Unsupervised curve clustering using B-splines. Scand. J. Stat. **3**(30), 581–595 (2003)
13. Coffey, N., Hinde, J., Holian, E.: Clustering longitudinal profiles using P-splines and mixed effects models applied to time-course gene expression data. Comput. Stat. Data Anal. **71**(3), 14–29 (2014)
14. de Boor, C.: A Practical Guide to Splines, Applied Mathematical Sciences Series. Springer, New York (2001)
15. Hastie, T.J., Tibshirani, R.J., Friedman, J.H.: The Elements of Statistical Learning: Data Mining, Inference, and Prediction. Springer, New York (2009)
16. Frasso, G., Eilers, P.H.C.: L- and V-curves for optimal smoothing. Stat. Model. **15**(1), 91–111 (2015)
17. Hansen, P.C.: Analysis of discrete ill-posed problems by means of the L-curve. SIAM Rev. **34**(4), 561–580 (1992)
18. Tibshirani, R., Walther, G., Hastie, T.: Estimating the number of clusters in a data set via the gap statistic. J. R. Stat. Soc. Ser. B (Stat. Methodol.) **63**(2), 411–423 (2001)
19. Arbeitman, M.N., Furlong, E.E.M., Imam, F., Johnson, E., Null, B.H., Baker, B.S., Krasnow, M.A., Scott, M.P., Davis, R.W., White, K.P.: Gene expression during the life cycle of drosophila melanogaster. Science **297**(5590), 2270–2275 (2002)
20. Penrose, L.S.: Distance, size and shape. Ann. Eugenics **17**(1), 337–343 (1952)
21. Hubert, L., Arabie, P.: Comparing partitions. J. Classif. **2**(1), 193–218 (1985)
22. Chiou, J.M., Li, P.L.: Functional clustering and identifying substructures of longitudinal data. J. R. Stat. Soc. Ser. B (Stat. Methodol.) **69**(4), 679–699 (2007)
23. Febrero-Bande, M., Oviedo de la Fuente, M.: Statistical computing in functional data analysis: the R package fda. usc. J. Stat. Softw. **51**(4), 1–28 (2012)
24. Kaufman, L., Rousseeuw, P.: Clustering by means of medoids. In: Dodge, Y. (ed.) Statistical Data Analysis Based on the L1 Norm and Related Methods, pp. 405–416. North Holland/Elsevier, Amsterdam (1987)
25. Lichman, M.: UCI Machine Learning Repository. University of California, Irvine, School of Information and Computer Sciences, http://archive.ics.uci.edu/m (2013)
26. Alcook, R.J., Manolopoulos, Y.: Time-series similarity queries employing a feature-based approach. In: Proceedings of the 7-th Hellenic Conference on Informatics, University of Ioannina (1999)
27. Jeong, Y.S., Jeong, M.K., Omitaomu, O.A.: Weighted dynamic time warping for time series classification. Pattern Recogn. **44**(9), 2231–2240 (2011)

# Comparing Multistep Ahead Forecasting Functions for Time Series Clustering

**Marcella Corduas and Giancarlo Ragozini**

**Abstract** The autoregressive metric between ARIMA processes has been originally introduced as the Euclidean distance between the AR weights of the one-step-ahead forecasting functions. This article proposes a novel distance criterion between time series that compares the corresponding multistep ahead forecasting functions and that relies on the direct method for model estimation. The proposed approach is complemented by a strategy for visual exploration and clustering based on the *DISTATIS* algorithm.

**Keywords** $AR$ metric · DISTATIS · Time series clustering · Multistep forecasting function

## 1 Introduction

A consolidated model-based approach to time series clustering sets up the problem in an inferential framework. Specifically, the dissimilarity between time series is measured by comparing the corresponding data generating mechanisms as described by independent Gaussian linear processes. In this context, the $AR$ metric measures the dissimilarity between two time series by means of the Euclidean distance between the weights of the $AR(\infty)$ formulation of each generating $ARIMA$ process [17]. The metric has been interestingly applied to several fields of analysis such as economics, environmental, and hydrological studies [8, 9, 16], and has been used to generalize the clustering approach to heteroskedastic series and to test causality relationships among time series [11, 14, 15].

However, in empirical studies, the classification produced by the $AR$ distance may be affected by the preliminary specification of the time series models. As a matter of fact, all statistical models are imperfect representation of reality, and the use of linear

M. Corduas (✉) · G. Ragozini
Department of Political Sciences, University of Naples Federico II, Naples, Italy
e-mail: marcella.corduas@unina.it

G. Ragozini
e-mail: giancarlo.ragozini@unina.it



F. Mola et al. (eds.), *Classification, (Big) Data Analysis and Statistical Learning*, Studies in Classification, Data Analysis, and Knowledge Organization,
https://doi.org/10.1007/978-3-319-55708-3_21

models belonging to a specific class inevitably makes them an approximation of the "true" data generating mechanism that could ignore certain dynamics. Uncorrect order identification, neglected long memory components, or nonlinearities are some of the situations that could affect the model identification step. For this reason, it is worth considering a possible extension of the above-mentioned metric in order to take model misspecification into account.

The present article is organized as follows. First, a novel distance measure based on the multistep ahead forecasting functions is discussed. Second, the *DISTATIS* algorithm is illustrated. Furthermore, some graphical tools for interpreting the results and a clustering strategy are presented. Finally, the performance of the proposed technique is verified with an empirical case study concerning the classification of hydrological time series.

## 2 The Distance Measure

Let $Z_t \sim ARIMA(p, d, q)$ be a zero mean invertible process such that

$$\phi(B)\nabla^d Z_t = \theta(B)a_t, \tag{1}$$

where $a_t$ is a Gaussian White Noise (WN) process with constant variance $\sigma^2$, $B$ is the backshift operator such that $B^k Z_t = Z_{t-k}, \forall k = 0, \pm 1, \ldots$, the polynomials $\phi(B) = 1 - \sum_{j=1}^{p} \phi_j B^j$ and $\theta(B) = 1 - \sum_{j=1}^{q} \theta_j B^j$, have no common factors, and all the roots of $\phi(B)\theta(B) = 0$ lie outside the unit circle.

The process $Z_t$ admits the $AR(\infty)$ representation:

$$\pi(B)Z_t = a_t, \tag{2}$$

where $\pi(B) = \phi(B)(1 - B)^d\theta^{-1}(B)$ being $\sum_{j=1}^{\infty} |\pi_j| < \infty$. Then, given a set of initial values, the $\pi$-weights sequence and the WN variance completely characterize $Z_t$, and a measure of structural diversity between two $ARIMA$ processes of known orders, $X_t$ and $Y_t$, is given by the Euclidean distance between the one-step-ahead forecasting functions:

$$d_{AR} = \sqrt{\sum_{j=1}^{\infty} \left(\pi_{xj} - \pi_{yj}\right)^2}. \tag{3}$$

A numerical approximation of $d_{AR}$ can be introduced by truncating the $\pi$-sequence at a suitable lag. The estimated distance is obtained from the maximum likelihood estimates of $ARIMA$ model parameters. Furthermore, the asymptotic distribution of the squared Euclidean distance has been derived under general assumptions about the model structure [7].

Note that the $AR$ distance is defined for invertible processes. Thus, the metric is applicable to both $ARMA$ processes as well as to $ARIMA$ and $ARFIMA$ processes. However, its application needs the preliminary identification of models from observed time series. This step can become rather complicated and time consuming when a large number of time series are involved. Thus, in practice, pure $AR$ models are often fitted to data using an automatic model selection criterion such as AIC or BIC [9]. This procedure may lead to a possible misspecification of the models. In such situations, it may be worth considering an alternative dissimilarity measure between two time series based on the comparison of the multistep ahead forecasting functions computed from the parameter estimates obtained by minimizing the sum of squares of in-sample $h$-step ahead forecast errors.

In particular, this estimation method has been widely investigated in literature, originating what is denoted as *direct* [6, 12] or *adaptive* estimation method [18]. The direct estimation approach implies that at each forecasting lead time, $h$, a different model is fitted to data. Specifically, the $h$-step forecasts for $h = 1, 2, \ldots, m$ are constructed by fitting the autoregression separately for each forecast period. When the orders of the model are not correctly identified, direct multistep estimation can lead to more efficient forecasts because local approximations are more flexible than global ones (see, for instance, [5]).

In this study, the $AR$ order of the model fitted to a given time series is held fixed for all the forecast periods and the $AR$ coefficients are estimated by minimizing:

$$\sum_t (Z_{t+h} \quad \phi_1^{(h)} Z_t - \cdots - \phi_{k_z}^{(h)} Z_{t-k_z+1})^2, \qquad h = 1, 2, 3, \ldots m. \tag{4}$$

The constraint concerning the $AR$ order $k_z$ can easily be removed. In such a case, at each forecast period, the order of the fitted model is identified by means of a modified formulation of the AIC or BIC criterion [3].

Moving from (4), we introduce a novel criterion for comparing time series, $X_t$ and $Y_t$, considering the set of estimated coefficients: $(\phi_{x1}^{(h)}, \ldots, \phi_{xk_x}^{(h)})$ and $(\phi_{y1}^{(h)}, \ldots, \phi_{yk_y}^{(h)})$ that are associated to the two time series at each lead time $h = 1, \ldots, m$. We define the multistep ahead distance between $X_t$ and $Y_t$ as the Euclidean distance:

$$\widetilde{d}_{(h)} = \sqrt{\sum_{j=1}^{L} \left(\phi_{xj}^{(h)} - \phi_{yj}^{(h)}\right)^2}, \quad h = 1, \ldots, m, \tag{5}$$

being $L = max(k_x, k_y)$ and having padded the vector of $AR$ coefficients with zeros when needed. The distance $\widetilde{d}_{(h)}$ will be zero when, given a set of initial values, both time series models produce the same direct $h-$step ahead forecast.

## 3 Visual Exploration and Time Series Clustering

Suppose that we have a collection of $r$ time series $(X_{1t}, \ldots, X_{rt})$, then we can construct the $r \times r$ distance matrix $\mathbf{D}_{(h)}$ for $h = 1, 2, \ldots, m$. We denote $\mathbf{D}_{(h)}$ as the $h$-step ahead distance matrix. In order to explore the global structure of this set of distance matrices, we propose to use the *DISTATIS* method [1, 2], that is, a generalization of the classical multidimensional scaling [10].

Although the Individual Differences Scaling (*INDSCAL*) and its modified versions are the most widely used algorithms [4, 13], for multiway analysis, the *INDSCAL* model theorizes the presence of a latent common structure underlying all the distance matrices and attempts to estimate such a common feature. Instead, *DISTATIS* is specifically designed with the aim of estimating a *compromise* that represents the best aggregation of the original distance matrices.

In particular, *DISTATIS* allows the user to visually explore the overall similarity or dissimilarity among the time series as measured by the multistep ahead distances. Furthermore, the method produces a useful synthesis of data, based on multistep ahead distances that can be the object of a subsequent clustering algorithm. Finally, it makes it possible to display, in the same graph, the position of each series according to the compromise (which accounts for all the multistep ahead distances at the same time) and according to the dissimilarity measured at a particular prediction period.

The main steps of the *DISTATIS* algorithm are the following. Given the distance matrices $\mathbf{D}_{(h)}$, $h = 1, \ldots, m$,

(1) Compute the cross-product matrices $\tilde{\mathbf{S}}_{(h)}$ according to the so-called "double-centering" transformation:

$$\tilde{\mathbf{S}}_{(h)} = -\frac{1}{2}\mathbf{C}\mathbf{D}_{(h)}\mathbf{C}^T,$$

with $\mathbf{C} = \mathbf{I} - r^{-1}\mathbf{1}\mathbf{1}^T$, where $\mathbf{I}$ is the $r$-dimensional identity matrix and $\mathbf{1}$ is the $r$-dimensional unit vector.

(2) Normalize each cross-product matrix $\tilde{\mathbf{S}}_{(h)}$:

$$\mathbf{S}_{(h)} = \gamma_{1(h)}^{-1}\tilde{\mathbf{S}}_{(h)},$$

where $\gamma_{1(h)}$ is the first eigenvalue of the matrix $\tilde{\mathbf{S}}_{(h)}$.

(3) Evaluate the similarity matrix between the cross-product matrices $\mathbf{A} = \{a_{h,h'}\}$, being

$$a_{h,h'} = \frac{\mathbf{s}_{(h)}^T\mathbf{s}_{(h')}}{||\mathbf{s}_{(h)}||||\mathbf{s}_{(h')}||}, \qquad h = 1, \ldots, m, \quad h' = 1, \ldots, m,$$

with $\mathbf{s}_{(h)} = vec(\mathbf{S}_{(h)})$. The matrix $\mathbf{A}$ is often indicated as the $RV$-matrix, because its elements are the $RV$ coefficients that measure the congruence between two matrices.

(4) Compute the eigenvalues and the eigenvectors of the matrix $\mathbf{A}$, i.e.,

$$\mathbf{A} = \mathbf{P}^T\boldsymbol{\Psi}\mathbf{P},$$

in order to obtain a factorial map representing all the considered distance matrices, each one as a point. This map is useful for the global assessment of the similarity among the $h$-steps ahead distance matrices.

(5) Compute the *compromise* matrix $\mathbb{S}$ as the weighted sum of the normalized cross-product matrices:

$$\mathbb{S} = \sum_{h=1}^{m} \alpha_h \mathbf{S}_{(h)}.$$

The weights $\alpha_h$ reflect the similarity among the distance matrices and are given by

$$\alpha_h = \frac{p_{h1}}{||\mathbf{p}_1||},$$

where $\mathbf{p}_1$ is the first eigenvector of the matrix $\mathbf{A}$. This eigenvector provides the contributions of the individual (normalized) cross-product matrices in determining the compromise. Moreover, $\psi_1 / \sum_{j=1}^{m} \psi_j$ measures the quality of the compromise.

(6) Perform the eigenvalue decomposition of the compromise matrix $\mathbb{S}$,

$$\mathbb{S} = \mathbf{V}\boldsymbol{\Lambda}\mathbf{V}^T$$

and compute the factorial coordinates to plot the data points (i.e., the time series) in the common space: $\mathbf{F} = \mathbf{V}\boldsymbol{\Lambda}^{\frac{1}{2}} = \mathbb{S}\mathbf{V}\boldsymbol{\Lambda}^{-\frac{1}{2}}$.

(7) Represent the cross-product matrices, $\mathbf{S}_{(h)}$, in the compromise space by projecting the matrices $\mathbf{S}_{(h)}$ as supplementary information. The coordinates can be easily computed as

$$\mathbf{F}_{(h)} = \mathbf{S}_{(h)}\mathbf{V}\boldsymbol{\Lambda}^{-\frac{1}{2}}.$$

Note that the compromise is a weighted average of the $h$-step ahead distance matrices using a double system of weights. The $\gamma_{1(h)}^{-1}$ coefficients express the relative importance of those distance matrices in terms of their inertia, whereas the $\alpha_h$ coefficients measure such importance with reference to the similarity between the distance matrices. Both weighting systems are data-driven and depend on the data structure.

## 4 A Case Study

The proposed technique has been applied to 20 time series consisting of 2700 observations of daily mean discharge (cubic feet per sec.) of unregulated rivers of Oregon and Washington state, as recorded from the U.S. Geological Survey. The considered data set is a useful benchmark since the dynamics of streamflow series may be characterized by long memory components which could be not well approximated by low-order $AR$ models.

Each observed time series has been preliminary transformed in logarithms, and in order to remove a strong deterministic seasonal pattern due to the climatic periodicity, the following regression model is fitted:

$$ln(W_t) = \sum_{j=1}^{4} \left[\beta_j \cos(2j\pi t/365.25) + \beta_j^* \sin(2j\pi t/365.25)\right] + Z_t.$$

This model is generally appropriate and seems to provide a useful summary of the stylized seasonal facts associated with the behavior of considered streamflow series [8]. Hence, the subsequent analysis is performed on estimated residuals from each regression model. The residual series conveys, in fact, significant information about the streamflow dynamics not depending on the succession of wet and dry periods. For each time series, the order of the $AR$ model has been selected according to the BIC criterion and the corresponding model has been estimated by means of the direct approach for $h = 1, 2, \ldots, 19$, assuming that we are interested in forecasting at most 19 days ahead. Thus, we have evaluated the distance matrices $\mathbf{D}_{(h)}$ for $h = 1, 2, \ldots, 19$.

Figure 1 shows the projection of each $h$-step ahead distance matrix, $\mathbf{D}_{(h)}$), onto the factorial space identified by the first and second eigenvectors of the $RV$-matrix.

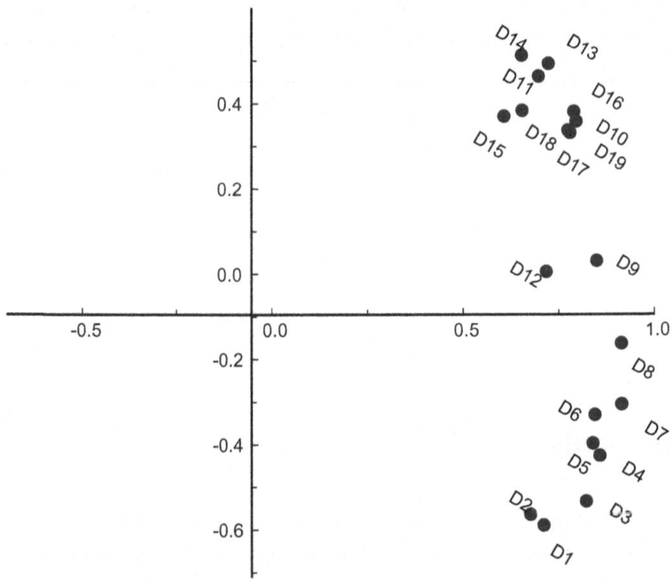


**Fig. 1** Plot of the between-studies space (i.e., eigen-analysis of the matrix $\mathbf{A}$; $\psi_1 = 11.38$, $\psi_2 = 3.02$)

The similarity between the distance matrices—computed at varying prediction horizons—can be immediately appreciated. All points have positive coordinates on the horizontal axis and they show some spread along the vertical axis. Moreover, very short lead times are opposite to long-term forecast periods. The coordinates on the first axis show that the $h$-step ahead distance matrices have similar weights $\alpha$s in the computation of the compromise matrix. The ratio of the first eigenvalue of the matrix $\mathbf{A}$ to the sum of its eigenvalues is 0.599: this implies that the compromise matrix explains about 60% of the inertia of the original set of distance matrices. Consequently, the $h$-step ahead distances differ on the information they capture about the time series. In addition, the common structure implied by such distances is reasonably summarized by the compromise matrix.

The points in Fig. 2 represent the objects, in other words the 20 time series, in the common space defined by the first two principal components derived from the compromise matrix. These components explain 44.5% of the inertia. The plot shows the presence of some clusters along with an element (12) that clearly appears far from the others. The hierarchical clustering using the complete linkage method identifies a partition of the series, as represented in the m-dimensional compromise space, into three groups and an isolated element: (1,2,7,11,17,19,20), (3,4,5,10,15,16,18), (6,8,9,13), and (12). Analogous clustering can be obtained from the coordinates produced by the classical MDS applied to the one-step-ahead distance matrix. In such a case, however, only two large clusters along with the isolated element are identified. The proposed technique appears, therefore, more accurate because it is able to discriminate better among models than the original $AR$ metric.

Furthermore, as mentioned previously, the *DISTATIS* method allows the comparison and representation of the series with respect to each forecasting period.

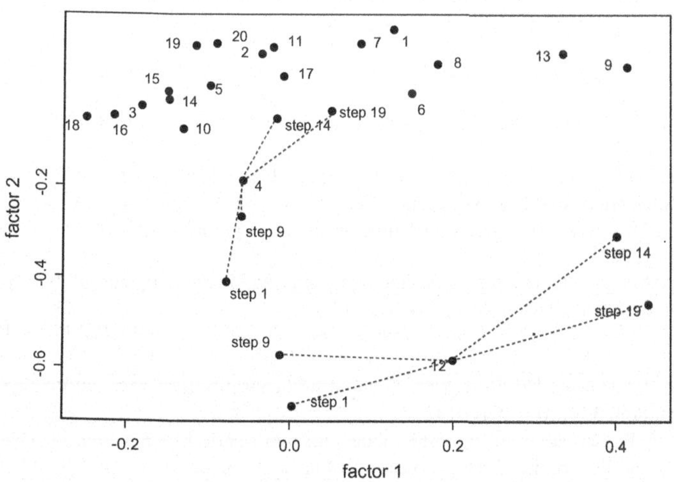


**Fig. 2** Time series in the compromise space and projections of the time series (4) and (12) at selected forecast horizons (1,9,14,19)

This is done by projecting the cross-product matrices onto the common spaces for $h = 1, \ldots, 19$. The position of a time series in the compromise is the barycenter of the 19 different positions that the time series has with reference to the considered multistep ahead distances matrices. In order to appreciate the variation of the $h$-step distances, in Fig. 2, we have plotted the projection of the outlying time series corresponding to a selection of forecasting periods, that is, 1, 9, 14, and 19 days. We have drawn segments linking the position of the time series (4) and (12) for each of the selected lead times to its compromise position. The series (12) is more sensitive to the differences between the forecasting time. This may be justified by the possible misspecification of the model. As a matter of fact, the $AR$ order selected by BIC is rather large compared to the other series and it could indicate the presence of neglected dynamic components.

In conclusion, this case study shows that the proposed approach is very promising. The $h$-step ahead distance measure seems to detect better the dissimilarity among time series models by enhancing the components that contribute to design the time series dynamics.

## References

1. Abdi, H., Valentin, D., O'Toole, A.J. Edelman, B.; DISTATIS: the analysis of multiple distance matrices. Proc.IEEE Computer Society: Computer Vision and Pattern Recognition, IEEE Computer Society, 42–47 (2005)
2. Abdi, H., Williams, L.J., Valentin, D., Bennani-Dosse, M.: STATIS and DISTATIS: optimum multitable principal component analysis and three way metric multidimensional scaling. Wiley Interdisc. Rev. Comput. Stat. **4**(2), 124–167 (2012)
3. Bhansali, R.J.: Asymptotically efficient autoregressive model selection for multistep prediction. Ann. Inst. Stat. Math. **48**, 577–602 (1996)
4. Carroll, J.D., Chang, J.J.: Analysis of individual differences in multidimensional scaling via an n-way generalization of Eckart-Young decomposition. Psychometrika **35**, 283–319 (1970)
5. Chevillon, G.: Direct multi-step estimation and forecasting. J. Econ. Surv. **21**, 746–785 (2007)
6. Clements, M.P., Hendry, D.F.: Multi-step estimation for forecasting. Oxf. Bull. Econ. Stat. **58**, 657–684 (1996)
7. Corduas, M.: La metrica autoregressiva tra modelli ARIMA: una procedura in linguaggio GAUSS. Quad. di stat. **2**, 1–37 (2000)
8. Corduas, M.: Clustering streamflow time series for regional classification. J. Hydrol. **407**, 73–80 (2011)
9. Corduas, M., Piccolo, D.: Time series clustering and classification by the autoregressive metric. Comput. Stat. Data Anal. **52**, 1860–1862 (2008)
10. Cox, T.F., Cox, M.A.: Multidimensional scaling. Chapman & Hall-CRC Press, Boca Raton (2000)
11. Di Iorio, F., Triacca, U.: Testing for Granger non-causality using the autoregressive metric. Econ. Modell. **33**, 120–125 (2013)
12. Findley, D.F.: On the use of multiple models for multi-period forecasting. In: Proceedings of Business and Economic Statistics, American Statistical Association, pp. 528–531 (1983)
13. Husson, F., Pagès, J.: `INDSCAL` model: geometrical interpretation and methodology. Comput. stat. Data Anal. **50**, 358–378 (2006)
14. Otranto, E.: Clustering heteroskedastic time series by model-based procedures. Comput. Stat. Data Anal. **52**, 4685–4698 (2008)

15. Otranto, E.: Identifying financial time series with similar dynamic conditional correlation. Comput. Stat. Data Anal. **54**, 1–15 (2010)
16. Palomba, G., Sarno, E., Zazzaro, A.: Testing similarities of short-run inflation dynamics among EU-25 countries after the Euro. Empirical Econ. **37**, 231–270 (2009)
17. Piccolo, D.: A distance measure for classifying ARIMA models. J. Time Ser. Anal. **11**, 153–164 (1990)
18. Tiao, G.C., Tsay, R.S.: Some advances in non-linear and adaptive modelling in time-series. J. Forecast. **13**, 109–131 (1994)

# Comparing Spatial and Spatio-temporal FPCA to Impute Large Continuous Gaps in Space

**Mariantonietta Ruggieri, Antonella Plaia and Francesca Di Salvo**

**Abstract** Multivariate spatio-temporal data analysis methods usually assume fairly complete data, while a number of gaps often occur along time or in space. In air quality data long gaps may be due to instrument malfunctions; moreover, not all the pollutants of interest are measured in all the monitoring stations of a network. In literature, many statistical methods have been proposed for imputing short sequences of missing values, but most of them are not valid when the fraction of missing values is high. Furthermore, the limitation of the methods commonly used consists in exploiting temporal only, or spatial only, correlation of the data. The objective of this paper is to provide an approach based on spatio-temporal functional principal component analysis (FPCA), exploiting simultaneously the spatial and temporal correlations for multivariate data, in order to provide an accurate imputation of missing values. At this aim, the methodology proposed in a previous proposal is applied, in order to obtain a good reconstruction of temporal/spatial series, especially in presence of long gap sequences, comparing spatial and spatio-temporal FPCA.

**Keywords** FDA · FPCA · GAM · P-splines

## 1 Introduction

Many statistical methods have been developed for missing data in low-dimensional settings, while methods of imputation in presence of a large number of variables, in high-dimensional settings, have not been systematically investigated. In air quality assessment, the presence of high percentages of missing values is often caused by

M. Ruggieri (✉) · A. Plaia · F. Di Salvo
Dipartimento di Scienze Economiche, Aziendali e Statistiche,
Università degli Studi di Palermo, Palermo, Italy
e-mail: mariantonietta.ruggieri@unipa.it

A. Plaia
e-mail: antonella.plaia@unipa.it

F. Di Salvo
e-mail: francesca.disalvo@unipa.it



F. Mola et al. (eds.), *Classification, (Big) Data Analysis and Statistical Learning*,
Studies in Classification, Data Analysis, and Knowledge Organization,
https://doi.org/10.1007/978-3-319-55708-3_22

system failures and technical issues with the complex monitoring equipment; when these failures occur, the data are not reported and gaps in time are generated. Usually, the networks are specifically designed to monitor a limited number of pollutants, and the other ones are monitored only at selected sites; in this case, we can observe the temporal dynamics of the pollutants in a restricted number of sites and the priority is integrating information that is misaligned in space and time. In this paper, the main issue is to find a suitable solution for a good reconstruction of long sequences of missing data in spatio-temporal series. Many existing methods dealing with missing values are not applicable or they do not exploit simultaneously both temporal and spatial correlation among data [8]. The functional data analysis (FDA) [11] can be considered as a method of imputation of missing values: observed time series are converted into a smoothed functional form, by reducing large amounts of data to few coefficients, without great loss of information, and the imputation of data is a consequence of the smoothing process. Nevertheless, FDA results a valid tool as a temporal gap-filling technique when short gaps occur.

In [13], temporal FPCA [11] is extended to multivariate case and employed [12] to impute very long gap sequences, providing a more accurate reconstruction.

In [2], we proposed spatial and spatio-temporal FPCA as approaches aiming at reducing random multivariate trajectories to a set of functional principal component scores: a three-mode eigenanalysis is conducted on the variance and covariance structures, which are estimated through the coefficients of suitable generalized additive models (GAMs); considering data as functions of space, or both time and space, the model reduces the observed information to a set of functional parameters, then the eigenanalysis synthesizes the dominant modes of their variation in the functional principal component. In this paper, we compare the ability of the proposed methodologies to estimate the functional data in presence of long sequences of missing values. As a result, we show that spatio-temporal FPCA takes advantage of both spatial and temporal correlations, outperforming spatial FPCA. The smoothing models are described in Sect. 2. Section 3 gives the variance function estimator and the representation of the functional data in terms of functional principal scores and weights. In Sect. 4, the proposed procedure is applied to air pollution data. The concentrations of five main pollutants ($PM$, $O_3$, $NO_2$, $CO$ and $SO_2$), recorded in 59 different monitoring stations in California in 2011, are standardized and scaled in [0, 100] according to [14] before performing any analysis, in order to account for different effects of each pollutant on human health, as well as for short- and long-term effects. Then, a simulation study is carried out and some performance indicators are computed to assess the effectiveness of the proposed procedure; results and comments are reported in Sect. 4. In Sect. 5, some possible further developments are discussed.

## 2 Methodology

The FDA [11] deals with modelling samples of random functions and, when data are spatially sampled curves, the interest has been recently focused on the spatial dependencies between curves. In estimating such functional data through smoothing model, a proper framework for incorporating space–time structures can be found in GAMs [5], while classical FPCA leads to eigenanalysis of the spatial covariance functions. GAMs are an extension of generalized linear models (GLMs) [10] in which the linear predictor is not restricted to be linear in the covariates, but it is the sum of smoothing functions applied to the covariates. GAMs are a good compromise between flexibility and complexity and provide a great tool to decide which model could be more adequate [5]. In our approach, we extend these ideas to the functional curves. There are some previous works in this direction, like the functional generalized linear models [1, 4] and the generalized spectral additive models [9]. According to FDA, the observed data $\boldsymbol{y}$, recorded at discrete times, are considered as realizations of a continuous process and we estimate functional data using tensor-product B-splines with roughness penalties, the penalized B-spline or P-spline approach detailed by [3]. Non-parametric methods are proposed in [7] to estimate correlation among functions with observations sampled at regular temporal grids. A flexible modelling methodology for spatio-temporal data is suggested by the generalized linear array model [6] applied to the smoothing of multidimensional arrays; the authors extend the P-spline approach in order to consider the smoothing over spatial and temporal dimensions, allowing both separable and non-separable structures. In a previous paper [2], we extend this approach to the multivariate and irregularly spaced time series, inside a general framework that links GAMs to multivariate functional data. This methodology has a considerable flexibility in modelling the dynamics of the functions through P-spline smoothing; here we consider two alternative models, the first one incorporating the spatial effects and the second one including separable space–time effects. In the first case, multivariate functional data are functions of the space and, for each of the $P$ dimensions, with $p = 1, \ldots, P$, we model the spatial effects on $\mathbf{y}^p$ considering the geographic coordinates $\mathbf{s} = (longitude, latitude)$ as explanatory variables:

$$\underbrace{y^p_{\mathbf{s}t}}_{data} = \underbrace{x^p_t(\mathbf{s})}_{signal} + \underbrace{\varepsilon^p_t(\mathbf{s})}_{noise}. \tag{1}$$

The random error $\varepsilon$ is assumed as normally distributed: $\varepsilon \sim \mathbf{N}(\mathbf{0}, \sigma^2\mathbf{I})$, although it may follow any exponential family distribution [5].

We assume a smooth function to model spatial effects in terms of linear combination of a basis matrix and $K$ coefficients:

$$x^p_t(\mathbf{s}) = \boldsymbol{\Phi}(\mathbf{s})\theta^p_t, \tag{2}$$

for each index $p$ in $(1, \ldots, P)$ and $t$ in $(1, \ldots, T)$. For irregularly spaced data, the matrix $\boldsymbol{\Phi} = (\boldsymbol{\Phi}_1(\mathbf{s})), \boldsymbol{\Phi}_2(\mathbf{s}), \ldots, \boldsymbol{\Phi}_K(\mathbf{s}))$ is the row-tensor product of the two marginal basis $\boldsymbol{\Phi}_1$ and $\boldsymbol{\Phi}_2$, one for each coordinate, with $k_1$ and $k_2$ being the number of parameter for each basis and $K = k_1 k_2$:

$$\boldsymbol{\Phi} = \boldsymbol{\Phi}_2 \square \boldsymbol{\Phi}_1 = (\boldsymbol{\Phi}_2 \otimes \mathbf{1}'_{\mathbf{k}_1}) \odot (\mathbf{1}'_{\mathbf{k}_2} \otimes \boldsymbol{\Phi}_1). \tag{3}$$

The symbol $\square$ indicates the box product, or row-wise Kronecker product, $\otimes$ is the Kronecker product and $\odot$ is the element-wise matrix product.

Using the $p-$dimensional data, $y_{\mathbf{s}}^{p}t$, observed in $S$ sites and $T$ time intervals and arranged in a $(S \times T \times P)$ array, the model (2) may be suitably represented in matrix terms as: $\mathbf{X}^p = \boldsymbol{\Phi}\Theta^p$ where, for each $p$, $\mathbf{X}^p$ is a $(S \times T)$ matrix, $\Theta^p$ is the $(K \times T)$ matrix of unknown parameters and $\boldsymbol{\Phi}$ is the $(S \times K)$ matrix of B-splines that spans the space of the covariate $\mathbf{s}$.

The second model generalizes the representations (1) and (2), in order to take into account space–time effects; in this case, a smooth three-dimensional function across space and time is considered:

$$\underbrace{y_{\mathbf{s}t}^{p}}_{data} = \underbrace{x^p(\mathbf{s}, t)}_{signal} + \underbrace{\varepsilon^p(\mathbf{s}, t)}_{noise}, \tag{4}$$

where

$$x^p(\mathbf{s}, t) = \Phi(\mathbf{s}, t)\theta^p. \tag{5}$$

In matrix terms, $\mathbf{X} = \Phi\Theta$, $\mathbf{X}$ being the functional data arranged in $ST$ rows and $P$ columns, $\Phi = \boldsymbol{\Phi}_s \otimes \boldsymbol{\Phi}_t$ is the spatio-temporal smoothing basis matrix with $ST$ rows and $K$ columns, $K = k_1\, k_2\, k_3$, and $\Theta$ is a matrix of coefficients $(K \times P)$. In both the models (2) and (5), the coefficients $\theta^p$ are estimated by minimizing the penalized residual sum of squares with a penalty matrix taking into account anisotropic smoothing structures, since the smoothness of each term is controlled by a single smoothing parameter:

$$PENSSE_{\lambda}(y) = \left\| \mathbf{y}_{s,t}^{p} - \boldsymbol{\Phi}\boldsymbol{\theta}^p \right\|^2 + \boldsymbol{\theta}'^p \mathbf{H} \boldsymbol{\theta}^p. \tag{6}$$

The penalty matrix $\mathbf{H}$ for the model (2) and for the model (5) is, respectively:

$$\mathbf{H} = \lambda_1 \mathbf{D_1}'\mathbf{D_1} \otimes \mathbf{I_{k_2}} + \lambda_2 \mathbf{I_{k_1}} \otimes \mathbf{D_2}'\mathbf{D_2}, \tag{7}$$

$$\mathbf{H} = \lambda_1 \mathbf{D_1}'\mathbf{D_1} \otimes \mathbf{I_{k_2}} \otimes \mathbf{I_{k_3}} + \lambda_2 \mathbf{I_{k_1}} \otimes \mathbf{D_2}'\mathbf{D_2} \otimes \mathbf{I_{k_3}} + \lambda_3 \mathbf{I_{k_1}} \otimes \mathbf{I_{k_2}} \otimes \mathbf{D_3}'\mathbf{D_3}. \tag{8}$$

The appropriate degree of smoothness can be estimated from data using different criteria such as AIC, BIC or GCV. More computational details about modelling multivariate data for spatio-temporal effects are reported in a previous paper [2].

## 3 Variance Functions

The dynamic of the spatial variability of the smoothed data can be quantified by the variance functions $\mathbf{V}^p(\mathbf{s}, \mathbf{s}^*) = COV(X^p(\mathbf{s}), X^p(\mathbf{s}^*))$, a suitable measure of the spatial information retained by the $p^{th}$ variable along time, as described in [11].

Due to the large number of variables and the different sources of variability, a synthesis of the variance structure may be achieved through a dimension reduction technique: the FPCA provides dimension reduction for functional data, finding the directions in the observation space along which the data have the highest variability, and determining uncorrelated linear combinations of the original variables. The eigenfunctions are determined by solving a functional eigenequation system:

$$\int\int v^{1,1}(\mathbf{s},t;\mathbf{s}^*,t^*)\xi^1(\mathbf{s},t)d\mathbf{s}^*dt^* + \cdots + \int v^{1,P}(\mathbf{s},t;\mathbf{s}^*,t^*)\xi^P(\mathbf{s},t)d\mathbf{s}^*dt^* = \rho\xi^1(\mathbf{s},t)$$
$$\ldots = \ldots$$
$$\int\int v^{p,1}(\mathbf{s},t;\mathbf{s}^*,t^*)\xi^1(\mathbf{s},t)d\mathbf{s}^*dt^* + \cdots + \int v^{p,P}(\mathbf{s},t;\mathbf{s}^*,t^*)\xi^P(\mathbf{s},t)d\mathbf{s}^*dt^* = \rho\xi^p(\mathbf{s},t)$$
$$\ldots = \ldots$$
$$\int\int v^{P,1}(\mathbf{s},t;\mathbf{s}^*,t^*)\xi^1(\mathbf{s},t)d\mathbf{s}^*dt^* + \cdots + \int v^{P,P}(\mathbf{s},t;\mathbf{s}^*,t^*)\xi^P(\mathbf{s},t)d\mathbf{s}^*dt^* = \rho\xi^P(\mathbf{s},t) \quad (9)$$

The theoretical derivation and the computational aspects of multidimensional (spatial and spatio-temporal) FPCA in the multivariate case are developed in [2]. The eigenfunctions, or principal component functions (FPCs), are orthonormal functions interpreted as the modes of variation and accounting for most of the variability expressed by the covariance functions; they also provide a comprehensive basis for representing the data, and hence, they are very useful in prediction of functional data. Here the FPCs are considered in an alternative perspective of gap-filling, to handle multivariate functions which contain large gaps in the temporal coverage or when, due to the sparse spatial coverage [15], missing data in many locations would lead to a substantial loss of information. The predictive performance between the spatial and spatio-temporal approach is compared through an application to pollution data and a simulation study.

### 3.1 *Reconstruction of Long Gaps in Functional Data*

We consider the reconstruction of functional data in presence of long gaps under the two functional models. Both the reconstructions are based on the empirical eigenfunctions of the covariance operators, but each of the eigendecompositions allows us to get a small dimension space which exhibits the main modes of variation under the respective functional model considered.
For the model (2), the solutions of the (9) are the eigenfunctions:

$$\xi_h^p(\mathbf{s}) = \Phi(\mathbf{s})\mathbf{b}_h^p. \quad (10)$$

The eigenfunctions for model (5) are computed as: $\xi_h(\mathbf{s},\mathbf{t}) = \Phi(\mathbf{s},t)\mathbf{b}^h$. The FPC scores $f^h(\mathrm{t}) = \sum_{p=1}^{P}\sum_{s=1}^{S} x_t^p(\mathbf{s})\xi_h^p(\mathbf{s})$ synthesize variability over space for the model (2). For the model (5), the FPC scores $f^h(p) = \sum_{t=1}^{T} \sum_{s=1}^{S} x_t^p(\mathbf{s})\xi_h(\mathbf{s}, \mathbf{t})$ synthesize variability over time and space.

After determining the number $H$ of eigenfunctions which sufficiently approximate the infinite dimensional process, the curves reconstruction based on the spatial model (2) is performed according to

$$\hat{x}_t^p(\mathbf{s}) = \mu(\mathbf{s}) + \sum_{h=1}^{H} f^h(\mathrm{t})\xi_h^p(\mathbf{s}), \tag{11}$$

while for the spatio-temporal smoothing model (5), the curves reconstruction is performed according to

$$\hat{x}^p(\mathbf{s}, t) = \mu(\mathbf{s}, t) + \sum_{h=1}^{H} f^h(p)\xi_h(\mathbf{s}, t). \tag{12}$$

## 4 An Application to Air Pollution Data and a Simulation Study

Missing data may occur for many reasons. For a given site, due to equipment failures and/or instruments errors, data can be missing along a time interval for all or some pollutants; otherwise, data can be missing by design, when the pollutant $p$ is not recorded at the station $\mathbf{s}$ (spatial misalignment).

In this section, a simulation study is carried out, in order to compare the reconstructions by spatial FPCA and spatio-temporal FPCA proposed in the previous section. In the simulation, we generate 100 missing data indicator arrays $\mathbf{M}$, from a Bernoulli distribution, with the same dimensions of the observed array. In each array $\mathbf{M}$ we randomly place long gaps along space, according to pollutant and time. Then for each model:

- The observed array $\mathbf{X}$ is converted into the functional $\tilde{\mathbf{X}}$.
- Each array $\mathbf{M}$ is laid upon $\mathbf{X}$ by creating 'artificial gaps' and obtaining $\mathbf{X}^{\mathbf{M}}$.
- Each array $\mathbf{X}^{\mathbf{M}}$ is converted into the functional $\tilde{\mathbf{X}}^{\mathbf{M}}$.
- FPCA is performed on $\tilde{\mathbf{X}}^{\mathbf{M}}$.

After extracting a number $F$ of FPCs from $\tilde{\mathbf{X}}^{\mathbf{M}}$, FPCs are used to reconstruct $\tilde{\mathbf{X}}^{\mathbf{M}}$ according to (11) and (12), obtaining a new array $\hat{\mathbf{X}}^{\mathbf{M}}$. Each $\hat{\mathbf{X}}^{\mathbf{M}}$ is compared with $\tilde{\mathbf{X}}$ by means of two performance indicators:

- *the correlation coefficient* $\rho$: $\rho = \left[\frac{1}{m}\frac{\sum_{l=1}^{m}[(\tilde{x}_l - M_{\tilde{x}})(\hat{x}_l - M_{\hat{x}})]}{\sigma_{\tilde{x}}\sigma_{\hat{x}}}\right]$ (the higher the better);

**Table 1** $\mu$ and $\sigma$ of performance indicators: data reconstructed by spatial (left) and spatio-temporal (right) FPCA

| | Spatial FPCA | | Spatio-temporal FPCA | |
|---|---|---|---|---|
| | (All missing) | (Long gaps) | (All missing) | (Long gaps) |
| $\mu_\rho$ | 0.9552 | 0.5310 | 0.9954 | 0.7510 |
| $\sigma_\rho$ | 0.0045 | 0.5071 | 0.0006 | 0.3885 |
| $\mu_{RMSD}$ | 4.6093 | 4.0534 | 1.4175 | 1.0018 |
| $\sigma_{RMSD}$ | 0.2076 | 2.7920 | 0.0865 | 0.7629 |

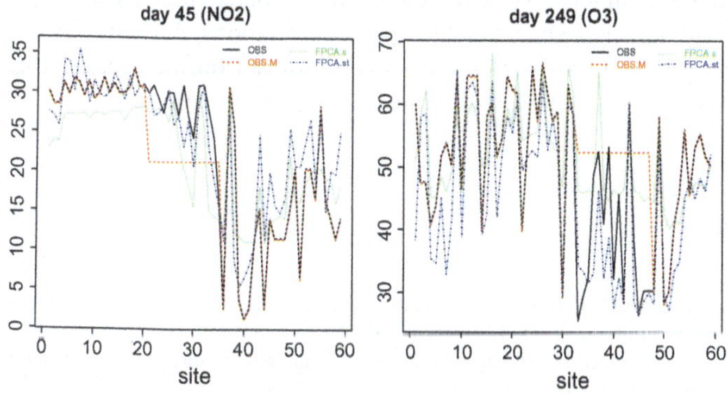


**Fig. 1** Long gaps reconstructed by spatial and spatio-temporal FPCA

- *the root mean square deviation RMSD*: $RMSD = \left(\frac{1}{m}\sum_{l=1}^{m}[\tilde{x}_l - \hat{x}_l]^2\right)^{1/2}$ (the lower the better).

where $m$ is the number of imputations; $\hat{x}_l$ is the $l^{th}$ imputed (considering space only or both space and time) data point, $l = 1, 2, \ldots, m$; $M_{\tilde{x}}$ ($M_{\hat{x}}$) and $\sigma_{\tilde{x}}$ ($\sigma_{\hat{x}}$) are the average and the standard deviation of all the $\tilde{x}_l$ ($\hat{x}_l$), respectively.

The distributions of $\rho$ and $RMSD$, over the 100 arrays $\mathbf{X}^{\mathbf{M}}$, are summarized by their means $\mu$ and standard deviations $\sigma$ by considering the whole set of missing values and only long gap sequences. Table 1 reports the obtained results for both the reconstruction by spatial and spatio-temporal FPCA. As it is evident, the space–time approach outperforms the space one, especially in presence of long gap sequences, by showing that exploiting simultaneously the temporal and the spatial correlations among data provides a more accurate reconstruction of data. In Fig. 1, an example of reconstruction for both space (green) and space-time (blue) approaches is also reported, as well as the observed data (black) and the observed data with gaps (red).

## 5 Discussion and Further Developments

In this paper, FPCA is suggested for dimension reduction applied to multivariate spatio-temporal data. GAMs or their generalizations may also have an important role to extend methodologies for FDA to the multivariate functional context.

Dealing with missing data, 100 arrays with long gaps are simulated and two alternative methodologies are applied: their performances are assessed by means of two performance indicators. Our results indicate that imputation of missing data based on spatio-temporal FPCA provides a much better reconstruction than spatial FPCA and, a gain in terms of accuracy and precision is obtained when we take into account simultaneously spatial and temporal correlations. A sensitivity analysis will be performed in a future work, to assess the impact on the proposed methods of different assumptions of the space–time correlation structures.

## References

1. Cardot, H., Sarda, P.: Estimation in generalized linear models for functional data via penalized likelihood. J. Multivar. Anal. **92**, 24–41 (2005)
2. Di Salvo, F., Ruggieri, M., Plaia, A.: Functional principal component analysis for multivariate multidimensional environmental data. Environ. Ecol. Stat. **22**(4), 739–757 (2015)
3. Eilers, P., Marx, B.: Flexible smoothing with B-splines and penalties. Stat. Sci. **11**, 89–121 (1996)
4. Escabias, M., Aguilera, A.M., Valderrama, M.J.: Principal component estimation of functional logistic regression discussion of two different approaches. J. Nonparametric Stat. **16**, 365–384 (2004)
5. Hastie, T., Tibshirani, R.: Generalized Additive Models. Chapman & Hall/CRC, Boca Raton (1990)
6. Lee, D., Durban, M.: P-spline ANOVA-type interaction models for spatio-temporal smoothing. Stat. Model. **11**, 49–69 (2011)
7. Li, Y., Wang, N., Hong, M., Turner, N.D., Lupton, J.R., Carroll, R.J.: Nonparametric estimation of correlation functions in longitudinal and spatial data, with application to colon carcinogenesis experiments. Ann. Stat. **35**(4), 1608–1643 (2007)
8. Liu, C., Ray, S., Hooker, G.: Functional Principal Components Analysis of Spatially Correlated Data, arXiv:1411.4681 (2014)
9. Muller, H.G., Yao, F.: Functional additive models. J. Am. Stat. Assoc. **103**, 1534–1544 (2008)
10. McCullagh, P., Nelder, J.: Generalized Linear Models, 2nd edn. Chapman & Hall/CRC, Boca Raton (1989)
11. Ramsay, J.O., Silverman, B.W.: Functional Data Analysis. 2nd edn. Springer-Verlag (2005)
12. Ruggieri, M., Di Salvo, F., Plaia, A., and Agró, G.: EOFs for gap filling in multivariate air quality data: a FDA approach. In: Lechevallier, Y. and Saporta G. (eds.), Proceedings of COMPSTAT 2010, Physica-Verlag, pp. 1557–1564 (2010)
13. Ruggieri, M., Di Salvo, F., Plaia, A., Agro', G.: Functional principal component analysis for the explorative analysis of multisite-multivariate air pollution time series with long gaps. J. Appl. Stat. **40**, 795–807 (2013)
14. Ruggieri, M., Plaia, A.: An aggregate AQI: comparing different standardizations and introducing a variability index. Sci. Total Environ. **420**, 263–272 (2012)
15. Yao, F., Muller, H., Wang, J.: Functional data analysis for sparse longitudinal data. J. Am. Stat. Assoc. **100**(470), 577–590 (2005)

# Part VII
# Finance and Economics

# A Graphical Tool for Copula Selection Based on Tail Dependence

Roberta Pappadà, Fabrizio Durante and Nicola Torelli

**Abstract** In many practical applications, the selection of copulas with a specific tail behaviour may allow to estimate properly the region of the distribution that is needed at most, especially in risk management procedures. Here, a graphical tool is presented in order to assist the decision maker in the selection of an appropriate model for the problem at hand. Such a tool provides valuable indications for a preliminary overview of the tail features of different copulas which may help in the choice of a parametric model. Its use is illustrated under various dependency scenarios.

**Keywords** Copula · Cluster analysis · Tail dependence · Graphical statistics Quantitative risk management

## 1 Introduction

In quantitative risk management, the main task is the study of the interdependencies among the involved random variables (or individual risk factors) moving away from simplifying assumptions (e.g. independence) and specific numerical quantities (e.g. linear correlation coefficients). This may allow to identify risks that seem to be highly unlikely to occur but could have a major impact on the estimation of the global distribution.

R. Pappadà (✉) · N. Torelli
Department of Economics, Business, Mathematics and Statistics,
University of Trieste, I-34127 Trieste, Italy
e-mail: rpappada@units.it

N. Torelli
e-mail: nicola.torelli@econ.units.it

F. Durante
Dipartimento di Scienze dell'Economia, Università del Salento,
I-73100 Lecce, Italy
e-mail: fabrizio.durante@unisalento.it



F. Mola et al. (eds.), *Classification, (Big) Data Analysis and Statistical Learning*,
Studies in Classification, Data Analysis, and Knowledge Organization,
https://doi.org/10.1007/978-3-319-55708-3_23

A risk model, in fact, can be represented by the probability distribution function of a multivariate random vector $\mathbf{X} = \{X_1, \ldots, X_d\}$, with continuous marginal distribution functions $F_1, \ldots F_d$ and can be expressed, by virtue of Sklar's Theorem [12], as

$$F(x_1, \ldots, x_d) = C(F_1(x_1), \ldots, F_d(x_d)), \tag{1}$$

where $C\colon [0,1]^d \to [0,1]$ is the copula of $\mathbf{X}$.

Thus, copulas are the functions that allow to aggregate individual risk factors into one global risk output and fully characterize the margin-free dependence structure of a random vector. Since their introduction, copulas have been largely employed in all areas of finance (see, e.g. [8]) as well as in environmental sciences (see, e.g. [11]) for the construction and estimation of multivariate stochastic models. In many practical applications, the risk associated with a given phenomenon is sensitive to the behaviour of the copula in specific regions of its domain, i.e. the tails of a distribution. For this reason, the determination of copulas with specific tail features may allow to estimate properly the region of the distribution that is needed at most for risk quantification (see, e.g. [3]). As such, there is a clear need for the further exploration of tools for the selection and validation of copula models especially when, due to data scarcity, classical goodness-of-it techniques may not be efficient. In [10], a graphical tool is applied for the investigation of goodness-of-fit of a set of parametric copula families by means of an Euclidean metric.

The aim of this work is to ease the consideration of a large number of possible copulas by means of suitable descriptive functions that can help distinguishing among different tail features. A copula graphical tool based on a dissimilarity providing different weights in the tails is introduced. It is shown how such tool can be used in the first steps of model building and provide valuable indications for the choice of a suitable model.

## 2 Graphical Tools to Detect Tail Dependence

As known, the concept of tail dependence is expressed by the conditional probability of one variable being extreme given that the other is extreme. In a bivariate copula setting, measures of tail dependence are designed to capture the dependence between the marginals in the upper-right quadrant and in the lower-left quadrant of $[0, 1]^2$. Traditional measures of tail dependence are the so-called *tail dependence coefficients* (shortly, TDCs), which give a measure calculated at the limit of the proportion of the probability mass of the joint distribution present in the tails (see [9]). It is important to stress that TDCs only depend on the diagonal section $\delta_C$ of the copula $C$ of the random pair $(X, Y)$ (see [5]) and are expressed (when they exist) as $\lambda_L = \lim_{t\to 0^+} \delta_C(t)/t$ and $\lambda_U = \lim_{t\to 1^-}(1 - 2t + \delta_C(t))/(1 - t)$, where $\delta_C(t) = C(t, t)$ for all $t \in [0, 1]$. Since they are asymptotic approximations of the tail behaviour of a copula, the TDCs actually concentrate on infinitely extreme co-movement. This gives them some unfortunate properties, such as an inability to discriminate between certain copulas.

One potential solution to this problem is to consider the so-called *tail concentration function* (TCF). Such a function well describes the features of different copulas since it represents a way to quantify the tail dependence at a finite scale (Fig. 1). The TCF for copula $C$ is the function $q_C : (0, 1) \to [0, 1]$ is given by

$$q_C(t) = \frac{\delta_C(t)}{t} \cdot \mathbf{1}_{(0,0.5]}(t) + \frac{1 - 2t + \delta_C(t)}{1 - t} \cdot \mathbf{1}_{(0.5,1)}(t). \tag{2}$$

For practical purposes, the tail concentration function can be more suited for measuring the extent to which pairs or groups of variables are linked at the extremes. In fact, Frank, Plackett and Normal copulas have zero upper (lower) tail dependence coefficients, but they exhibit a different behaviour at finite scale, as shown in Fig. 1 (left panel). The right panel of Fig. 1 displays the tail concentration function for copulas with non-zero lower (upper) tail dependence.

Here, the idea is to exploit such an information in order to distinguish among different candidates in the choice of a parametric model on the basis of the lower or upper tail behaviour, and provide a graphical tool to assess which copulas more closely capture specific features of the data. To this end, we proceed according to the following steps:

1. Define the copula test space, i.e. the set of all possible copula models that may be appropriate for the data at hand.
2. As done also in [10], consider a suitable metric between the empirical copula (as derived from data) and the (fitted) parametric copulas in the test space. Here, the key information is represented by the tail concentration function approximated from data.
3. Finally, define a suitable dissimilarity matrix and graphically visualize the goodness-of-fit for pairs of random variables.

Specifically, let $(X_i, Y_i)_{i=1,\ldots,n}$ be a bivariate sample from an unknown copula and consider a set of $k$ copulas $C_1, C_2, \ldots, C_k$ belonging to different families that have

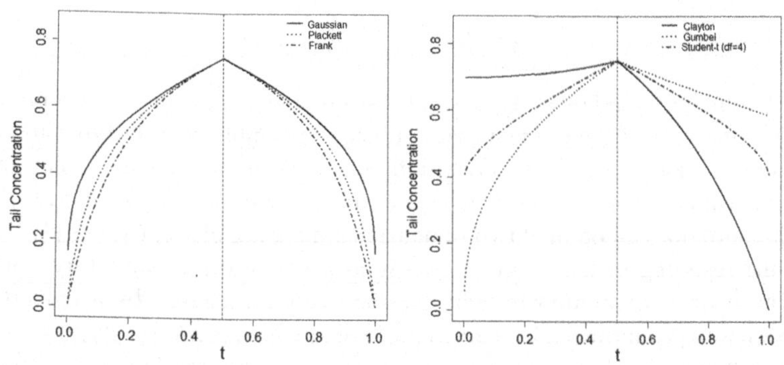


**Fig. 1** Tail concentration functions for popular families of copulas with the same Blomquist's $\beta$ coefficient

been fitted to the available data. As known, the empirical copula $C_n$ is defined, for all $(u, v) \in [0, 1]^2$, by

$$C_n(u, v) = \frac{1}{n} \sum_{i=1}^{n} \mathbf{1}(U_i \le u, V_i \le v), \tag{3}$$

where, for all $i = 1, \ldots, n$, $U_i = R_i/(n+1)$ and $V_i = S_i/(n+1)$, $R_i$ denoting the rank of $X_i$ among $X_1 \ldots, X_n$ and $S_i$ the rank of $Y_i$ among $Y_1 \ldots, Y_n$. A dissimilarity between the empirical copula $C_n$ and the copula $C_i$, $i = 1, \ldots, k$, can be defined as

$$\sigma(C_n, C_i) = \int_a^b w(t) \left(q_{C_n}(t) - q_{C_i}(t)\right)^2 dt, \tag{4}$$

for suitable $a$ and $b$ with $0 \le a < b \le 1$, where $w\colon [a, b] \to [0, +\infty]$ is a weight function, $q_{C_n}$ is the TCF associated with the empirical copula in Eq. (3) and $q_{C_i}$ is the TCF associated with $C_i$, $i = 1, \ldots, k$. Analogously, the dissimilarity between the $i$-th and the $j$-th copula is

$$\sigma(C_i, C_j) = \int_a^b w(t) \left(q_{C_i}(t) - q_{C_j}(t)\right)^2 dt, \tag{5}$$

for $1 \le i, j \le k, i \ne j$. The quantities in Eqs. (4 and 5) can be seen as generalization of the dissimilarity measures introduced in [3], where both lower and upper parts of the TCF are considered. A risk-weighted approach is adopted here, which can be useful when the interest is in investigating the strength of similarity among $C_n, C_1, \ldots, C_k$ at very low or very high quantiles, as is the case of the application presented in Sect. 3. A dissimilarity matrix $\Delta = (d_{ij})$ of order $(k+1)$ is then defined, whose elements are

$$\begin{aligned} d_{1j} &= \sigma(C_n, C_{j-1}), & j &= 2, \ldots, k+1, \\ d_{ij} &= \sigma(C_{i-1}, C_{j-1}), & i, j &= 2, \ldots, k+1, \quad i < j \\ d_{ii} &= 0, & i &= 1, \ldots, k+1. \end{aligned} \tag{6}$$

Thus, $\Delta$ contains a kind of $L^2$-type distances computed by means of an approximation on a fine grid of the domain that is related to sample size. In order to obtain a low-dimensional representation respecting the ranking of distances, a non-metric multidimensional scaling (MDS) technique can be performed on $\Delta$. As a result, the procedure provides an optimal two-dimensional mapping of $C_n$, $C_i$, $i \in \{1, \ldots, k\}$, where the resulting distances fit as closely as possible the dissimilarity information. The final configuration is such that the distortion caused by a reduction in dimensionality is minimized by a given criterion. Finally, the $K$ points $p_i = (x_i, y_i)$ corresponding to copula $C_i$ and $p_{emp} = (x_{emp}, y_{emp})$ corresponding to the empirical copula $C_n$ can be visualized in a 2D graph.

It is worth stressing that other proxies can be used in order to detect finite tail dependence. For instance, one can consider the *Spearman's tail concentration*, given by $s_L(t) = \rho(X, Y|X \leq t, Y \leq t)$ for $t \in (0, 0.5]$ and $s_U(t) = \rho(X, Y|X \geq t, Y \geq t)$ for $t \in [0.5, 1)$, where $\rho$ denotes the Spearman's rank correlation coefficient. This measure has been used, for instance, by [4]. Another possibility could be to introduce a tail concentration function with respect to the secondary diagonal of the copula domain, particularly useful in the presence of negative dependence.

These graphical copula tools based on different tail dependence measures can be used to shed light on the problem of classifying copulas according to some desirable characteristics of practical relevance, especially when tail dependence properties are of primary interest. For instance, they may supply valuable indications for a preliminary analysis of the extremal joint behaviour of the positions in a portfolio, so that a more informed model choice can be made. This latter aspect may play an important role in financial applications for identifying assets that tend to be highly associated under certain market conditions, i.e. for diversification purposes.

In this contribution, we focus on bivariate copulas but the concept of tail concentration function can be generalized to the multivariate case. To this end, the work by [2] could be applied to define a suitable concept of TCF in a multivariate setting.

## 3 An Application to Financial Time Series

To illustrate the proposed tool, we consider Morgan Stanley Capital International (MSCI) Developed Markets indices, designed to measure the equity market performance of 23 developed markets. The data set consists of $d = 23$ time series ($T = 2094$ daily observations) from June 4, 2002 to June 10, 2010, including the following countries: Australia, Austria, Belgium, Canada, Denmark, Finland, France, Germany, Greece, Hong Kong, Ireland, Italy, Japan, Netherlands, New Zealand, Norway, Portugal, Singapore, Spain, Sweden, Switzerland, the United Kingdom and the United States. A first step of data filtering is applied, as common in the literature, where ARMA-GARCH models with $t$-distributed innovations are fitted to the univariate time series of log returns, and the standardized residuals $\widehat{z_{it}}, t \in \{1, \ldots, T\}$ are computed for each $i \in \{1, \ldots, d\}$. For all $i, j \in \{1, \ldots, d\}$, $i < j$, define the vector of *pseudo-observations* $\tilde{\mathbf{U}}_t^{ij} = (\tilde{U}_{i,t}, \tilde{U}_{j,t})$ via probability integral transformation

$$\tilde{U}_{i,t} := \frac{T}{T+1} F_{i,T}(\widehat{z_{it}}), \quad \tilde{U}_{j,t} := \frac{T}{T+1} F_{j,T}(\widehat{z_{jt}}), \tag{7}$$

for $t \in \{1, \ldots, T\}$, where $F_{i,T}$ and $F_{j,T}$ denote the empirical distribution functions of the $i$-th and the $j$-th margin, respectively. The dependence among the time series is hence fully expressed by the copula of the estimated probability integral transform variables (concentrated in $[0, 1]^d$) obtained after these fittings (see [4]).

Now, following the procedure described in Sect. 2, we consider the set of one-parameter copulas $\mathcal{C} = \{\mathcal{C}^1, \dots, \mathcal{C}^K\}$ and fit a copula $C_k$ from the family $\mathcal{C}^k$ to $\tilde{\mathbf{U}}_t^{ij}$, $k = 1, \dots, K$. The copula parameter is estimated via inversion of Kendall's $\tau$ (if the copula family has more than one parameter, a maximum likelihood procedure can be adopted). Being interested in modelling the lower tail dependence among the time series, we can consider the dissimilarities in Eqs. (4 and 5) with $a = 0$, $b = 1$ and $w(t) = \mathbf{1}_{(0,0.5]}(t)$, that is, we only consider the (unweighted) lower part of Eq. (2). The left panel of Fig. 2 displays, for instance, the empirical tail concentration function for time series New Zealand and Hong Kong. By considering all the copulas in the test space, the $K$ dissimilarities between $C_i$ and the empirical copula associated with pseudo-observations from Eq. (7) are computed, as well as the $K(K-1)/2$ mutual dissimilarities between the $i$-th and the $j$-th fitted copula. Specifically, we consider the following copula models: Clayton, Gumbel, Frank, Normal, Joe, Plackett, Galambos, Student's $t$ with 4 and 8 degrees of freedom, Survival Gumbel, Survival Clayton and Survival Joe. This defines the $13 \times 13$ dissimilarity matrix $\Delta$. Finally, the non-metric MDS algorithm takes all the dissimilarities and assigns each one to a location in a low-dimensional space, by minimizing the so-called *stress function*, expressed in terms of the inter-points distances in the new configuration and some monotonic transformation of the elements of $\Delta$.

The procedure discussed above provides a graphical representation of the empirical copula and the 12 fitted copulas in two dimensions (for a stress value lower than 2.5%), for all possible pairs of time series in the data set. The right side of Fig. 2 displays the graphical tool for New Zealand and Hong Kong. The data clearly do not exhibit greater dependence in the positive tail than in the negative, and our tool suggests that the Survival Gumbel and the Student-$t$ copula with 4 degrees of

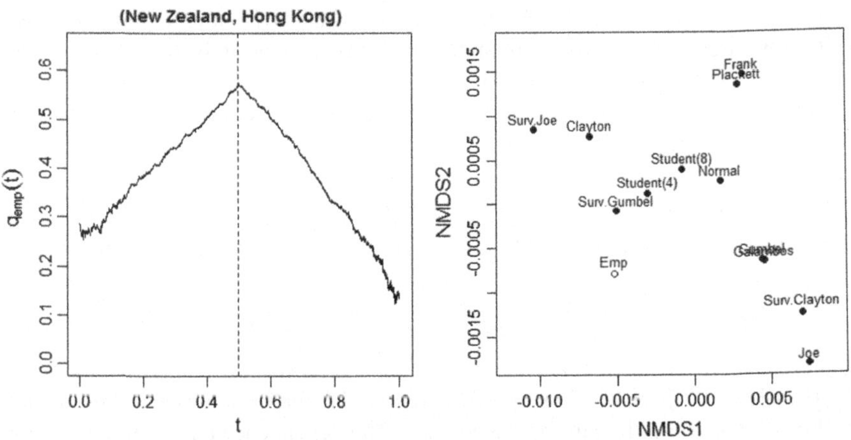


**Fig. 2** Empirical TCF (left) and two-dimensional representation of goodness-of-fit for time series New Zealand–Hong Kong from the MSCI Developed Markets Index, based on the weight function $w(t) = \mathbf{1}_{(0,0.5]}(t)$ (right). The estimated value of Kendall's tau is 0.164

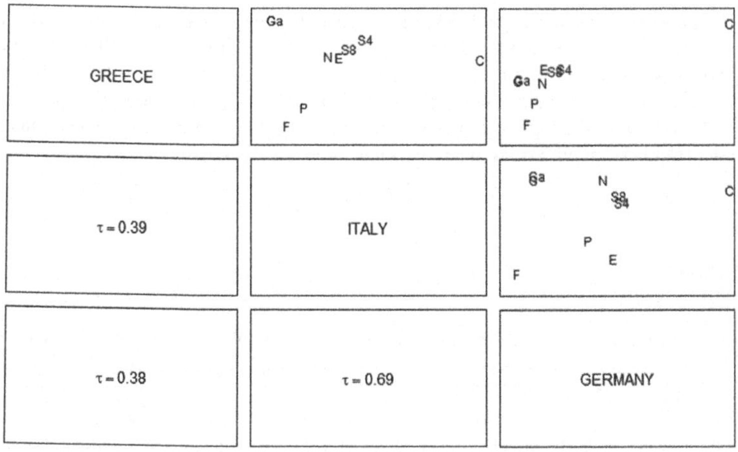


**Fig. 3** Pairs plots for three MSCI time series. Lower panels display the values of Kendall' tau. Upper panels display the 2D goodness-of-fit representation, where the empirical copula is denoted with letter 'E' and the copulas in the test space are Normal (N), Frank (F), Plackett (P), Clayton (C), Gumbel (G), Galambos (Ga), Student-$t$ with 4 (S4) and 8 (S8) degrees of freedom

freedom can be considered suitable possibilities since they appear to be closer to $p_{emp}(=\texttt{Emp})$, the empirical copula. The two models both allow for joint fat tails and seem to be appropriate for describing the data.

By displaying all the $d(d-1)/2$ pairwise 2D goodness-of-fit representations for the entire data set of $d$ time series, the proposed graphical tool can provide a complete overview of the copulas to choose among at a bivariate level, and then classical techniques (see, e.g. [1, 6, 7]) can be used to validate that choice. Figure 3 shows the application of the described tool to daily returns of time series Greece, Italy and Germany, and a copula test space of cardinality 8.

## 4 Conclusions

Copulas provide a flexible way to model correlated variates. Several copulas with specific features are available for modelling such dependencies. A graphical tool is introduced to illustrate different properties that can distinguish among several copula families. It is based on some descriptive functions that can be used to select copulas having desired characteristics (i.e. tail concentration) and to judge how well the fitted copulas match specific aspects of the data. A suitable copula-based dissimilarity measure is introduced, and an illustration from MSCI data shows how the presented tool can provide useful guidelines for modelling multivariate data sets which can exhibit complex patterns of dependence in the tails.

**Acknowledgements** The second author has been partially supported by the Faculty of Economics and Management (Free University of Bozen-Bolzano, Italy), via the project 'NEW-DEMO'. The other authors acknowledge the support of the University of Trieste, FRA 2014 ('Metodi e modelli matematici e statistici per la valutazione e gestione del rischio in ambito finanziario e assicurativo') and FRA 2016 ('Nuovi sviluppi di statistica e matematica applicata per la previsione, l'analisi e la gestione dei rischi con applicazioni in ambito finanziario e assicurativo').

## References

1. Choroś, B., Ibragimov, R., Permiakova, E.: Copula estimation. In: Jaworski, P., Durante, F., Härdle, W.K., Rychlik, T. (eds.) Copula Theory and Its Applications. Lecture Notes in Statistics, vol. 198, pp. 77–91. Springer (2010)
2. De Luca, G., Rivieccio, G.: Multivariate tail dependence coefficients for Archimedean copulae. In: Di Ciaccio, A., Coli, M., Angulo, I., Jose M. (eds.) Advanced Statistical Methods for the Analysis of Large Data-Sets, pp. 287–296. Springer, Berlin (2012)
3. Durante, F., Fernández-Sánchez, J., Pappadà, R.: Copulas, diagonals, and tail dependence. Fuzzy Sets Syst. **264**, 22–41 (2015)
4. Durante, F., Pappadà, R., Torelli, N.: Clustering of financial time series in risky scenarios. Adv. Data Anal. Classif. **8**, 359–376 (2014)
5. Durante, F., Sempi, C.: Principles of Copula Theory. CRC/Chapman & Hall, Boca Raton, FL (2016)
6. Fermanian, J.D.: An overview of the goodness-of-fit test problem for copulas. In: Jaworski, P., Durante, F., Härdle, W.K. (eds.) Copulae in Mathematical and Quantitative Finance. Lecture Notes in Statistics, pp. 61–89. Springer, Berlin (2013)
7. Genest, C., Rémillard, B., Beaudoin, D.: Goodness-of-fit tests for copulas: a review and a power study. Insurance Math. Econom. **44**(2), 199–213 (2009)
8. Jaworski, P., Durante, F., Härdle, W.K. (eds.): Copulae in Mathematical and Quantitative Finance. Lecture Notes in Statistics, vol. 213. Springer, Berlin (2009)
9. Joe, H.: Parametric families of multivariate distributions with given margins. J. Multivar. Anal. **46**(2), 262–282 (2009)
10. Michiels, F., De Schepper, A.: A new graphical tool for copula selection. J. Comp. Graph. Stat. **22**(2), 471–493 (2013)
11. Salvadori, G., De Michele, C., Kottegoda, N., Rosso, R.: Extremes in Nature: an Approach to Using Copulas, Water Science and Technology Library, vol. 56. Springer, Berlin (2007)
12. Sklar, A.: Fonctions de répartition à *n* dimensions et leurs marges. Publications de l'Institut Statistique de l'Universite de Paris **8**, 229–231 (1959)

# Bayesian Networks for Financial Market Signals Detection

**Alessandro Greppi, Maria E. De Giuli, Claudia Tarantola and Dennis M. Montagna**

**Abstract** In order to model and explain the dynamics of the market, different types and sources of information should be taken into account. We propose to use a Bayesian network as a quantitative financial tool for market signals detection. We combine and incorporate in the model, accounting, market, and sentiment data. The network is used to describe the relationships among the examined variables in an immediate way. Furthermore, it permits to identify in a mouse-click time scenario that could lead to operative signals. An application to the analysis of S&P 500 index is presented.

**Keywords** Bayesian network · Index analysis · Market signals detection

## 1 Introduction

A large amount of information could be used to elucidate the market behavior. For institutional and retail investors, it can be difficult to interpret and quickly elaborate the available information in order to identify operative signals. Since market signals detection involves probabilistic reasoning under uncertainty, we propose to employ Bayesian networks (BNs).

We use BNs to develop a comprehensive model incorporating accounting, market, and sentiment data. A BN, see, e.g., [4] and [2], is a tool for modeling multivariate probability structures and making inference. It is a probabilistic graphical model that represents a set of random variables and their conditional dependencies via a directed

A. Greppi (✉) · M. E. De Giuli · C. Tarantola · D. M. Montagna
Department of Economics and Management, University of Pavia, Pavia, Italy
e-mail: alessandro.greppi01@universitadipavia.it

M. E. De Giuli
e-mail: elena.degiuli@unipv.it

C. Tarantola
e-mail: ctaranto@unipv.it

D. M. Montagna
e-mail: dennis.montagna@unipv.it

 
F. Mola et al. (eds.), *Classification, (Big) Data Analysis and Statistical Learning*,
Studies in Classification, Data Analysis, and Knowledge Organization,
https://doi.org/10.1007/978-3-319-55708-3_24

acyclic graph (DAG). The nodes of the DAG correspond to the variables under investigation, and the joint probability distribution is expressed in terms of the product of the conditional tables associated to each node given its parents. The use of a graph as a pictorial representation of the problem at hand simplifies model interpretation and facilitates communication and interaction among investors with different kinds and degrees of information. BNs allow the user to identify unknown and/or unexpected relationships among variables of different areas. Efficient information propagation techniques can be exploited to assess the effect of changes in the status of specific variables on the remaining ones. Hence, BNs can be a useful instrument to support institutional and retail investors in their decision processes. In this paper, we present an application of BNs to the analysis of the S&P 500 index.

The structure of the paper is the following. In Sect. 2, we present the examined data. Sections 3 and 4 are devoted to the application of BNs to market signals detection. Finally, in Sect. 5, we end up with some final remarks.

## 2 Data Description

We examine the behavior of the S&P 500 index for the period from January 1, 2004 to October 23, 2015. We use weekly data downloaded from the "Bloomberg" database. We construct and apply a BN to identify the variables that influence "buy/sell" signals for the S&P 500 index. There is no agreement in the financial community regarding which indicators and variables should be considered to this aim. We follow the recommendation of Credit Suisse [8] that suggests considering the following categories of variables: value, growth, profitability, sentiment, momentum, and technical analysis. The variables of these categories provide relevant quantitative and qualitative information for market index analysis.

In addition, we consider the sixth category including contrarian variables. These variables provide indications against the price trend of an index. When the market is performing well, they give a sell signal in order to reduce market exposure; on the other hand, if the market is performing poorly, they provide a buy signal. For more details on the use of contrarian variables, see, e.g., [5]. We consider the following two contrarian variables: B_S_CRB and B_S_SPX. The first one is related to the price of the Commodities Research Bureau (CRB) index; the second one is related to the price of the S&P 500 index.

We list below the examined variables; for a detailed description of these variables, see the supplementary material of the "Bloomberg" database.

1. *Value*: Price\Earnings (PE_RATIO), Price\Sales (P_SALES), Price\Cash Flow (P_CF), Ebitda per Share (EBITDA_PS), Sales per Share (SALES_PS), EV\Ebitda (EV_EBITDA), EV\Sales (EV_SALES), Price\Book Value (P_BV), and Dividend Yield (DVD_YLD)
2. *Growth*: 3Y Moving Average Sales Growth (SALES_GR), 3Y Moving Average Ebitda Growth (EBITDA_GR), and 3Y Moving Average Earnings Growth (EARN_GR)

3. *Profitability*: Ebitda Margin (EBITDA_MRG) and Buy-Back Yield (BB_YLD)
4. *Sentiment*: Volatility (VOLA) and Put\Call Ratio (PC_RATIO)
5. *Momentum and Technical Analysis*: Index value above or below the mean market price calculated over the previous 52 weeks (P_UP_DOWN), Rate of Change (ROC), and Relative Strength Index (RSI)
6. *Contrarian Variables*: Buy/Sell the Commodities Research Bureau (CRB) index (B_S_CRB), and Buy/Sell the S&P 500 (B_S_SPX) index

It is a common practice of practitioners to work with a discretized version of the previous variables. We consider the following classification.

Variable P_UP_DOWN assumes two states: 0 when the value of the S&P 500 index is lower than its one-year average, 1 otherwise.

Variables of categories 1–5 are classified in three states. The considered coding scheme is based on how the values of each examined variable differ from its median ($me$). We assign label 1 to all values greater than $\left(me + \frac{1}{4}\sigma\right)$ (where $\sigma$ is the standard deviation); we assign label 2 to all values smaller than $\left(me - \frac{1}{4}\sigma\right)$; and we assign label 0 to all remaining values.

For the contrarian variables, the labeling is inverted. More precisely, state 2 corresponds to a high value (sell signal), while state 1 represents a low value (buy signal). State 0 does not provide a clear operative indication.

## 3 The Financial Market Network

The structure of the network is learned directly from the available data by using the Hugin software (www.hugin.com). We consider the following two steps procedure:

(1) In order to solve sparsity problems, we run the Chow-Liu algorithm to draw an initial draft of the network, see [1]. This algorithm is used to obtain a tree structure maximizing the data likelihood. Logical constraints, such as target variables having no outgoing arrows toward any of the other variables, have been taken into account.
(2) We run the NPC (Necessary Path Condition) algorithm, see [9], using the tree obtained in step 1 as an initial restriction. We impose additional constraints deriving from our financial market knowledge. Using the preliminary procedure described in step 1, we reduce the dimensionality of the model space to explore with the NPC algorithm. Finally, the NPC algorithm allows us to choose, among independence equivalent models, the most suitable one for the problem at hand.
The construction of the model is completed via the estimation of the conditional probability tables (the parameters of the BN) from the data; this is achieved via the EM (Expectation-Maximization) algorithm whose version for BNs has been proposed by [6]. The EM algorithm consists of two steps: the expectation step (E-step) and the maximization step (M-step). In the E-step, we calculate the expected data frequencies given the current value of the parameters; in the M-step, we maximize the log-likelihood of the parameters under the expected data frequencies. These two steps are alternated iteratively until a predetermined stopping criterion is satisfied.

Figure 1 represents the learnt dependence structure together with the marginal probability tables estimated from the data. By moving through the network, we can identify which are the variables that influence directly and indirectly the behavior of the output variable (B_S_SPX). The node representing the volatility of the S&P 500 index (VOLA) plays an important role in the BN. It transfers the information collected from its neighbors to the output node. According to [7], market volatility is a good risk measure for short-term investments. The Put\Call Ratio (PC_RATIO) is connected directly and indirectly to B_S_SPX. A value above 1 indicates negative expectations (bearish market); on the other hand, a value below 1 reflects positive expectations (bullish market). The node corresponding to the Relative Strength Index (RSI) is connected directly and indirectly (through the VOLA node) to the output node. This index compares the magnitude of recent gains to recent losses. If the market is overbought, it leads to a sell signal; when the market is oversold, we obtain a buy signal. As expected, the CRB Index (B_S_CRB) node influences directly the B_S_SPX, transferring the information collected from its neighbors to the output node. Finally, differently from common financial knowledge, Price\Earnings (PE_RATIO) is only indirectly related to B_S_SPX. More importantly, it only mildly influences the output variable (see Sect. 4.2).

## 4 Examination of Different Scenarios

Once the model has been estimated, the network can be used to study the effects of alternative market configurations. Different scenarios can be examined by inserting and propagating the appropriate evidence through the network. The effect of a change on a variable (or set of variables) can be obtained in a mouse-click time by the evidence propagation algorithm. Inference is conducted via message passing in a junction-tree representation of the network. More precisely, the junction tree is a particular hypergraph where nodes are grouped in cliques and separators. It is updated in two stages. First, the root updates its probability table collecting evidence from all the other cliques of the tree, and then the new evidence is distributed from the root to the leaves. For more details, see, e.g., [4] and [2].

In the following paragraphs, we present the results referring to the effects of a volatility shock and a change of the P\E ratio on the remaining variables of the network. For the analysis of other scenarios, see [3].

### *4.1 Scenario A: The Effect of Volatility*

In this scenario, we explore the effect of a sudden modification in the volatility level. We present how the conditional probability tables vary after the introduction of new information in the network. We compare the probability values associated to each state before (see Fig. 1) and after new evidence propagation (see Fig. 2).

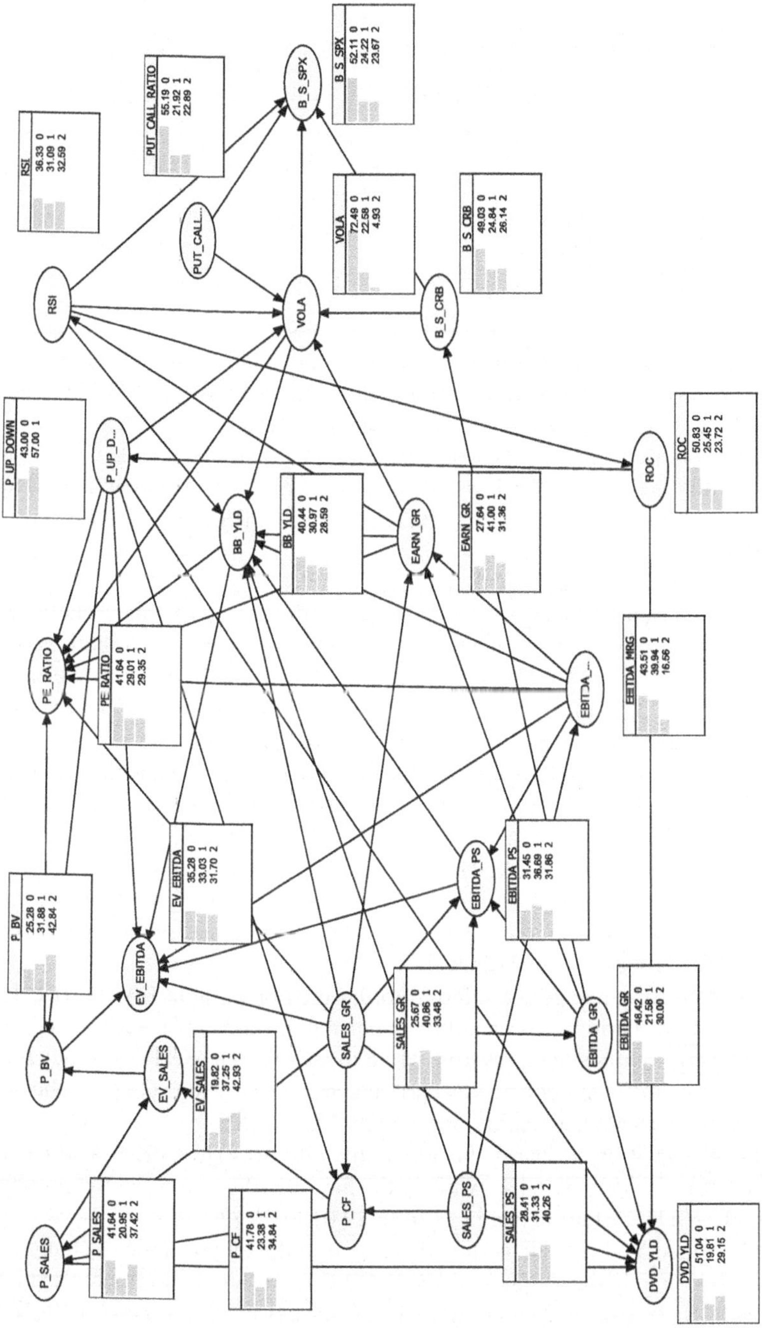


**Fig. 1** The BN for the S&P 500 index

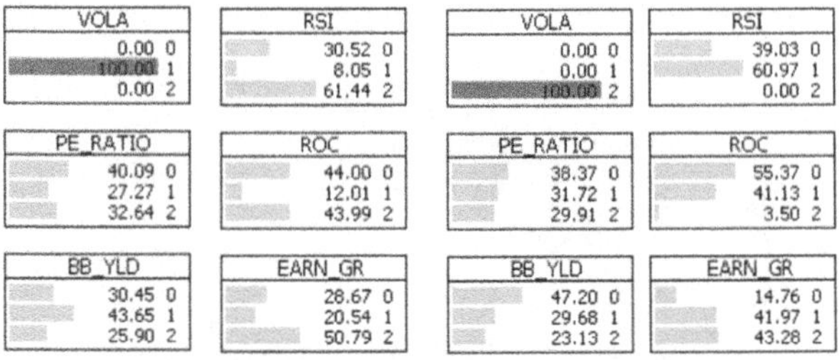


(a) High volatility (b) Low Volatility

**Fig. 2** The rule of volatility

We simulate two extreme situations: the case in which volatility assumes always the highest level (high volatility scenario), and the case in which volatility assumes always the lowest level (low volatility scenario). We insert this evidence in the network by double clicking on the corresponding states of node VOLA (state 1 for high volatility and state 2 for low volatility). In Fig. 2a, b, we show the effect of high/low volatility on variables PE_RATIO, BB_YLD, RSI, ROC, EARN_GR, and B_S_SPX.

The momentum and technical analysis variables, RSI and ROC, are the most sensible to a volatility change. This is in line with common financial knowledge. In the high volatility scenario, the probability of being in state 2 for both variables increases substantially (for RSI the probability rises from 32.59 to 61.44%, and it increases from 23.71 to 43.99% for ROC). In the low volatility scenario, we can observe a substantial increase in the probability associated to state 1.

On the profitability side, we can notice that BB_YLD is rather sensible to an increase in volatility. In the high volatility scenario, the probability referred to a high value of BB_YLD (state 1) increases from 30.97 to 43.65%. In the low volatility scenario, prices are more stable. Hence, companies can decide to slow down their repurchase activity, waiting for future buy opportunities. In fact, we notice that probabilities associated to states 1 and 2 of BB_YLD both decrease, while the probability of state 0 rises from 40.44 to 47.20%.

Other interesting results are referred to the growth variable EARN_GR. In the high volatility scenario, the probability associated to state 2 of EARN_GR (growth below the expectations) increases from 31.36 to 50.79%. The S&P 500 index is more subject to sell-offs when companies earnings are not growing. On the other hand, low volatility does not necessarily imply high EARN_GR.

In the high volatility scenario, the probability referred to low PE_RATIO (state 2) increases from 29.35 to 32.64%, while in the low volatility scenario, the probability of high PE_RATIO (state 1) increases from 29.01 to 31.72%.

Finally, since VOLA node is directly related to the output node, a change in its value affects directly the output node B_S_SPX. As expected, in the high volatility scenario, B_S_SPX shows a buy signal, while in the low volatility scenario, B_S_SPX provides a sell signal. For more details, see [3].

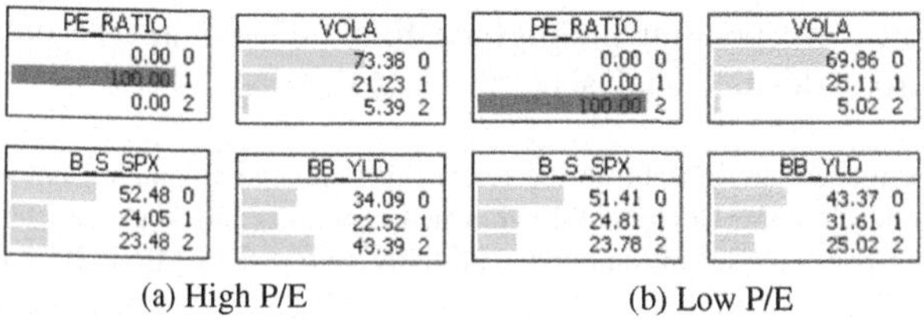


**Fig. 3** The rule of price\earnings

## 4.2 *Scenario B: The Effect of Price\Earnings*

Investors and practitioners consider PE_RATIO as the key variable for market index evaluation. In contrast with the common financial belief, a change in the PE_RATIO affects in a sensible way only the profitability variable BB_YLD. We represent in Fig. 3 the most interesting findings. The probability tables associated to each state before inserting new evidence are the ones reported in Fig. 1.

The impact of PE_RATIO on BB_YLD confirms that companies repurchase their own shares according to their P/E multiple. In the high PE_RATIO scenario (Fig. 3a), the probability referred to low BB_YLD increases from 28.59 to 43.39%, showing that the buy-back slows down when the market is "expensive". On the other hand, when the market is "cheap", the probability referred to high BB_YLD increases from 30.97 to 31.61%, see Fig. 3b.

Comparing the marginal probability tables in Fig. 3 with the ones in Fig. 1, it is clearly evident that a change in the value of PE_RATIO does not relevantly affect volatility and the buy/sell signal for S&P 500 index.

# 5 Concluding Remarks

In this paper, we illustrated how BNs can be a fruitful instrument for market signals detection. BNs provide a pictorial representation of the dependence structure between the examined variables, allowing us to discover dependencies that are difficult to reveal via common tools usually employed by financial operators. We proposed a comprehensive model, based on accounting, market, and sentiment variables to analyze the behavior of the S&P 500 index. Through a visual inspection of the network, it is possible to identify variables that directly or indirectly affect the buy/sell signal for the index. Furthermore, the network can be used to evaluate in real time how new information affects market dynamics and to compare the effectiveness of alternative financial strategies. We also applied BNs to the analysis of the market behavior in the period 1994–2003. This period was characterized by some of the deepest financial crisis (for example, Southeast Asia and Dot-Com Bubble crises). As expected given the peculiar characteristics of the examined period, the learnt structure and the corresponding findings are different from the ones presented in this paper; see [3] for a detailed analysis of this crisis period.

**Acknowledgements** We are grateful to the referee for valuable comments and suggestions.The first author is grateful to the Doctoral Research in Economics and Management of Technology (DREAMT). The work of the second author was partially supported by MIUR, Italy, PRIN MISURA 2010RHAHPL.

## References

1. Chow, C.K., Liu, C.N.: Approximating Discrete Probability distributions with dependence trees. IEEE Trans. Inf. Theor. **14**, 462–467 (1968)
2. Cowell, R.G., Dawid, A.P., Lauritzen, S.L., Spiegelhalter, D.J.: Probabilistic Networks and Expert Systems. Springer, New York (1999)
3. Greppi, A.: Bayesian Networks Models for Equity Market, Ph.D. Thesis, University of Pavia (2016)
4. Jensen, F.V.: Bayesian Networks. UCL press, London (1996)
5. Lakonishok, J., Shleifer, A., Vishny, R.W.: Contrarian Investment, Extrapolation, and Risk. J. Finance **49**(5), 1541–1578 (1994)
6. Lauritzen, S.L.: The EM Algorithm for Graphical Association Models with Missing Data. Comp. Stat. & Data Anal. **19**, 191–201 (1995)
7. Nielsen, A.E.: Goal - Global Strategy Paper No. 1, Goldman Sachs Global Economics - Commodities and Strategy Research. http://www.goldmansachs.com/our-thinking/archive/, (2011)
8. Patel, P.N., Yao, S., Carlson, R., Banerji, A., Handelman, J.: Quantitative Research - A Disciplined Approach, Credit Suisse Equity Research (2011)
9. Steck, H.: Constraint-Based Structural Learning in Bayesian Networks Using Finite Data, Ph.D. thesis, Institut für Informatik der Technischen Universität München (2001)

# A Multilevel Heckman Model to Investigate Financial Assets Among Older People in Europe

**Omar Paccagnella and Chiara Dal Bianco**

**Abstract** This paper applies a multilevel Heckman model to investigate the household behaviour (in an ageing population) on portfolio choices across Europe. Exploiting the richness of the data collected by the Survey of Health, Ageing and Retirement in Europe, both the ownership pattern and the amount invested in short-term and long-term assets are analysed. This statistical solution is suitable to take into account both the hierarchical nature of the data and the features of the variables of interest. Model estimates support the choice of a multilevel framework. The sample selection approach allows to highlight different results when analysing the ownership of a financial product rather than the amount invested in it.

**Keywords** Ageing · Financial assets · Multilevel modelling · Sample selection

## 1 Introduction

Heckman sample selection models are often applied when the variable of interest is observed or recorded only if a (selection) condition applies [10]. This may occur because of unit or item (survey) non-responses, or because a specific product or condition is owned by a subsample of units, and unobservable components affecting the inclusion in the subsample are correlated with unobservable factors influencing the variable of interest. The latter case may be well represented by the amount invested in financial and/or real assets: it may be observed and studied only if the unit has included that product in its own portfolio.

O. Paccagnella (✉)
Department of Statistical Sciences, University of Padua,
via C. Battisti, 241-35121 Padova, Italy
e-mail: omar.paccagnella@unipd.it

C. Dal Bianco
Department of Economics and Management, University of Padua,
via del Santo, 33-35123 Padova, Italy
e-mail: chiara.dalbianco@unipd.it



F. Mola et al. (eds.), *Classification, (Big) Data Analysis and Statistical Learning*,
Studies in Classification, Data Analysis, and Knowledge Organization,
https://doi.org/10.1007/978-3-319-55708-3_25

Household portfolios have been widely investigated, particularly among the old people [11]. As income of older people typically reflects pension income, consumption later in life is generally sustained by spending down financial or real assets [5]. Therefore, wealth is a key measure of the individual socio-economic status in an ageing society: older individuals control a substantial part of the household wealth, even though they usually have restricted portfolio holdings and do not invest in risky financial assets [9]. In the literature, this topic has been addressed by investigating separately the ownership patterns of financial products and the amount invested in the products [4, 13].

This paper aims at shedding more light on the behaviour of the households among the older population *across Europe*, studying *both* their ownership patterns and their invested amount in some financial assets (i.e. short-term or long-term investments). To this aim, in order to take into account the hierarchical nature of the data (households nested into countries) and the features of the variable of interest (the amount invested in a financial product is observed only if the household owns that asset), a multilevel Heckman model is the suggested methodological solution for the analysis. Such model is applied to data from SHARE (Survey of Health, Ageing and Retirement in Europe), which is an international survey on ageing, collecting detailed information on the socio-economic status of the European old population [2].

This paper is organised as follows: Sect. 2 introduces the statistical solution adopted for the analysis, while Sect. 3 briefly describes data. In Sect. 4, the main results and findings of the analysis are presented. Some concluding remarks and suggestions for future research are provided in Sect. 5.

## 2 The Model

Whenever sample selection is present, inferences based on standard regression analyses may lead to biased and inconsistent results. This phenomenon is much more complicated in a hierarchical setting because (i) the selection process may affect different levels of the hierarchy; (ii) the estimation procedure is more complex. See [7] for a discussion of the consequences of sample selection in multilevel or mixed models, particularly for the variance components.

According to [14], let $j = 1, ..., J$ be the level-2 units and $i = 1, ..., n_j$ be the level-1 units: the multilevel Heckman model is defined by two equations on two latent variables, the regression component:

$$y^*_{ij} = z'_{ij}\beta + u_{1j} + \varepsilon_{1ij} \tag{1}$$

and the selection component:

$$h^*_{ij} = w'_{ij}\gamma + u_{2j} + \varepsilon_{2ij}. \tag{2}$$

The observed responses are

$$y_{ij} = y^*_{ij},\ h_{ij} = 1\ if\ h^*_{ij} > 0$$

$$y_{ij}\ not\ observed,\ h_{ij} = 0\ if\ h^*_{ij} \leq 0, \tag{3}$$

where $z_{ij}$ and $w_{ij}$ are vectors of explanatory variables (they may contain the same elements, even if it is a good practice to specify at least one exclusion restriction), while $\beta$ and $\gamma$ are vectors of parameters to be estimated. The model is therefore characterised by two level-1 error terms ($\varepsilon_{1ij}$ and $\varepsilon_{2ij}$) and two level-2 error terms ($u_{1j}$ and $u_{2j}$). Error terms at the same level are assumed to be correlated, while error terms at different levels are uncorrelated.

In order to estimate the model by maximum likelihood through the *gllamm* (generalised linear latent and mixed models) procedure of Stata software [15], the correlation between level-1 error terms is induced by means of a shared random effect $\upsilon_{ij}$:

$$\begin{aligned}
\varepsilon_{1ij} &= \lambda\upsilon_{ij} + \tau_{ij} \\
\varepsilon_{2ij} &= \upsilon_{ij} + \omega_{ij} \\
\upsilon_{ij} &\sim N(0, 1) \\
\tau_{ij} &\sim N(0, \sigma^2_\tau) \\
\omega_{ij} &\sim N(0, 1),
\end{aligned} \tag{4}$$

where $\upsilon_{ij}$, $\tau_{ij}$ and $\omega_{ij}$ are independent each others. Level-1 covariance matrix and correlation term are equal, respectively, to

$$V(\varepsilon_{1ij}, \varepsilon_{2ij}) = \Sigma_1 = \begin{bmatrix} \lambda^2 + \sigma^2_\tau & \lambda \\ \lambda & 2 \end{bmatrix} \tag{5}$$

$$\rho = \frac{\lambda}{\sqrt{2\left(\lambda^2 + \sigma^2_\tau\right)}}. \tag{6}$$

Level-2 error terms are normally distributed, with zero mean, variance $\sigma^2_{u1}$ and $\sigma^2_{u2}$, respectively, and covariance $\sigma_{u12}$. Intraclass correlation coefficients (ICCs) can be defined both for the regression and the selection equation:

$$ICC^R = \frac{\sigma^2_{u1}}{\sigma^2_{u1} + (\lambda^2 + \sigma^2_\tau)}$$

$$ICC^S = \frac{\sigma^2_{u2}}{\sigma^2_{u2} + 2}. \tag{7}$$

## 3 Data

This paper uses data from the 2010–2011 wave (the fourth) of SHARE [1] (DOIs: https://doi.org/10.6103/SHARE.w4.500). See [2] for methodological details, and [3, 12] for more information on SHARE wave 4.

SHARE is an interdisciplinary survey on ageing that is run every 2 years in a host of European countries. It collects extensive information on health, socio-economic status and family interactions of people aged 50 and over.

Our sample includes 40129 households living in 16 countries (Austria, Belgium, Czech Republic, Denmark, Estonia, France, Germany, Hungary, Italy, the Netherlands, Poland, Portugal, Slovenia, Spain, Sweden and Switzerland). Within each household, we chose the eligible reference person ('head') as the respondent who holds the largest individual earnings.

Overall, household heads are equally divided between males and females (51% vs. 49%, respectively), even though male respondents are predominant in Germany and Spain (more than 60%), while their proportion is lower than 40% in Estonia. The mean age is equal to 67 years, on the whole: across countries, the median age ranges from 65 to 70 years. The household size is equal to one in about 30% of the households, while 60% of household heads report to live with a partner; the average household size ranges from 1.75 (Sweden) to 2.77 (Poland). Data show a large cross-country heterogeneity also according to education: high education (a degree or more) characterises more than 40% of household heads in Denmark, Germany and Sweden, while Mediterranean countries (Italy, Portugal and Spain) have the largest proportions (more than half of the sample) of low educated respondents. In Estonia, Hungary, Italy, Slovenia and Spain more than 80% of the households live in their accommodation as owners; in Austria, Germany and Switzerland, the proportion of non-owner households is about 40%, instead.

The variables of interest are the amount invested by the household in two derived financial product categories: (i) government/corporate bonds, stocks and mutual funds; (ii) individual retirement accounts, contractual saving and whole life insurance. In the SHARE questionnaire, the ownership and the amount invested in each financial product are asked separately, but imputed and cleaned values are provided for the two aggregated categories only. According to [8], the first category comprises different (medium and high) risky financial products with a short/middle time span; the second one groups long-term investments. As shown in Table 1, there is evidence of a large cross-country heterogeneity both in the ownership and the amount held in these assets.

## 4 Empirical Application

The multilevel Heckman model, introduced in Sect. 2, is estimated separately for short-term and long-term investments. Each model is regressed on a wide set of

**Table 1** Ownership and amount invested in the two asset categories. Country averages

| Country | Bond, stock & mutual funds | | Long-term investments | |
|---|---|---|---|---|
| | Proportion of owners (%) | Average amount among owners | Proportion of owners (%) | Average amount among owners |
| Austria | 15.4 | 42294.19 | 54.2 | 18283.11 |
| Belgium | 32.7 | 83881.07 | 44.3 | 30671.37 |
| Czech Republic | 9.8 | 10128.08 | 53.0 | 10968.69 |
| Denmark | 51.2 | 33818.76 | 42.3 | 81258.25 |
| Estonia | 5.2 | 19417.70 | 67.6 | 2309.94 |
| France | 24.3 | 28134.27 | 63.7 | 54787.33 |
| Germany | 34.8 | 46126.84 | 44.6 | 31381.24 |
| Hungary | 6.5 | 15799.00 | 18.7 | 13278.79 |
| Italy | 23.6 | 40531.93 | 6.6 | 35778.65 |
| The Netherlands[a] | 24.0 | 46033.96 | – | – |
| Poland | 2.4 | 6425.71 | 27.8 | 5894.59 |
| Portugal | 8.7 | 35054.43 | 26.2 | 52551.83 |
| Slovenia | 19.1 | 13436.96 | 15.1 | 13801.54 |
| Spain | 8.0 | 40455.50 | 14.4 | 43410.76 |
| Sweden | 67.5 | 34251.97 | 60.0 | 34655.04 |
| Switzerland | 42.8 | 107907.90 | 37.8 | 76289.09 |

Amounts are PPP-adjusted

[a]Excluded from the long-term investments' analysis because of an error (not yet corrected at the time of writing the paper) in one component of the assets

covariates, covering demographics (gender, age, having a partner and number of living children), socio-economic status (education, occupation and social network—defined as the number of individuals who make up the social network of the respondent), health (self-reported health according to the US scale, from 1-Excellent to 5-Very bad), cognitive abilities (according to the numeracy test, from 1-Very low to 5-Very high) and wealth (yearly amount of household real assets and income). All economic and financial variables are transformed using the inverse hyperbolic sine.

Although the set of covariates specified in the selection equation may be the same introduced in the regression equation, it is a good practice to specify at least one exclusion restriction in the selection equation in order to avoid problems of weak identification of the model. As suggested by [6], promising restrictions are interviewer or interviewing fieldwork characteristics. Interviewer characteristics are not available in the public data set, while interview information are included in the SHARE interviewer module. In this analysis, two exclusion restrictions are exploited: the presence of bothering factors during the interview (a dummy variable) and the respondent willingness to answer to the questions (an index from 1-Very good to 4-Very bad).

Results are reported in Table 2. Because of the presence of missing values in some covariates, the number of observations used in the selection equation is slightly lower than the total number of households.

In both analyses, correlations among level-1 error terms are positive and statistically significant, while correlations among level-2 error terms are negative and statistically significant only for long-term investments. This supports the adoption of a two-step Heckman approach.

Moreover, it is worth noting the different behaviour of ICCs among the two financial assets' categories: while for bonds, stocks and mutual funds cross-country

**Table 2** Model estimations, separately for short-term and long-term investments

| Variable | Short-term investments | | Long-term investments | |
|---|---|---|---|---|
| | Selection equation | Regression equation | Selection equation | Regression equation |
| Male | 0.109 *** | 0.144 *** | 0.040 * | 0.159 *** |
| Age | 0.106 *** | 0.256 *** | −0.300 *** | −0.373 *** |
| Having partner | −0.017 | −0.278 *** | 0.057 ** | 0.930 *** |
| Number living children | −0.123 *** | −0.054 *** | −0.041 *** | −0.047 *** |
| Middle education | 0.390 *** | 0.173 *** | 0.174 *** | 0.166 *** |
| High education | 0.772 *** | 0.327 *** | 0.430 *** | 0.457 *** |
| Self-reported health | −0.142 *** | −0.108 *** | −0.032 *** | −0.112 *** |
| Numeracy | 0.185 *** | 0.127 *** | 0.042 *** | 0.140 *** |
| Worker | −0.045 | −0.121 ** | 0.550 *** | 0.457 *** |
| Unemployed/disabled | −0.200 *** | 0.012 | 0.078 | 0.083 |
| Homemaker/other | −0.017 | 0.152 * | 0.113 ** | 0.535 *** |
| Social network | 0.089 *** | 0.008 | 0.071 *** | 0.059 *** |
| Household real assets | 0.080 *** | 0.052 *** | 0.032 *** | 0.028 *** |
| Household income | 0.478 *** | 0.543 *** | 0.161 *** | 0.205 *** |
| Bothering factors | 0.039 | – | 0.024 | – |
| Willingness to answer | −0.034 | – | −0.101 *** | – |
| Constant | −8.375 *** | 1.312 ** | −0.360 ** | 6.903 *** |
| *Level-1* | | | | |
| Variance | 2.000 | 2.236 | 2.000 | 2.836 |
| Covariance between errors | 0.485 ** | | 1.319 *** | |
| Correlation between errors | 0.229 | | 0.554 | |
| *Level-2* | | | | |
| Variance | 0.207 | 0.141 | 0.469 | 0.755 |
| Covariance between errors | −0.002 | | −0.102 *** | |
| Correlation between errors | −0.013 | | −0.171 | |
| ICC | 9.4% | 5.9% | 19.0% | 21.0% |
| Sample size | 38685 | 8136 | 38685 | 17617 |
| *Testing non-multilevel versus multilevel structure* | | | | |
| LR statistics ($\chi^2_3$) | 3049.37 *** | | 19950.77 *** | |

Significance levels: *** = 1%; ** = 5%; * = 10%

variability is quite low, particularly in the regression equation, the opposite applies for long-term investments. However, LR test for comparing a multilevel versus a non-multilevel structure strongly results in favour of the former for each financial asset.

The demographic characteristics seem to be more related to the amount invested in the asset, rather than to the decision of investing. Ceteris paribus, older people are more likely to invest in short-term assets, while the opposite relationship appears for long-term investments. Having a partner is negatively correlated with the amount invested in short-term assets, while it is positively correlated with long-term ones. Compared to the estimates of other covariates, occupational status shows a weaker association with both the decision to invest and the invested amount. On the contrary, wealth measures and education are strongly positively correlated with the dependent variables. Cognitive abilities seem to play an important role in the choice to participate in the financial market, as well as being in bad health.

## 5 Conclusions

The multilevel Heckman model is a good solution for investigating the behaviour of the households on portfolio choices across Europe. On the one hand, the multilevel approach may be able to take into account cultural or financial market differences across countries. On the other hand, a sample selection approach allows to exploit a (two-step) statistical solution to analyse both the ownership pattern of an asset and the amount invested in it.

A limitation of the present work could be the small number of countries that may affect the estimation of level-2 variances. Level-1 sample size is quite large and each model estimation shows a wide set of statistically significant variables. However, the occupational status is weakly related to the analysed financial assets, particularly for the short-term investments.

The literature on the multilevel Heckman model is currently very limited. A suggestion for future works on this topic is to investigate the introduction of some variables at the country level, in order to further explain the role of the country in the household portfolio choices.

**Acknowledgements** The SHARE data collection has been primarily funded by the European Commission through FP5 (QLK6-CT-2001-00360), FP6 (SHARE-I3: RII-CT-2006-062193, COMPARE: CIT5-CT-2005-028857, SHARELIFE: CIT4-CT-2006-028812) and FP7 (SHARE-PREP: N.211909, SHARE-LEAP: N.227822, SHARE M4: N.261982). Additional funding from the German Ministry of Education and Research, the U.S. National Institute on Aging (U01_AG09740-13S2, P01_AG005842, P01_AG08291, P30_AG12815, R21_AG025169, Y1-AG-4553-01, IAG_BSR06-11, OGHA_04-064) and from various national funding sources is gratefully acknowledged (see www.share-project.org).

## References


1. Börsch-Supan, A.: Survey of Health, Ageing and Retirement in Europe (SHARE) Wave 4. Release version: 5.0.0. SHARE-ERIC. Data set. (2016). https://doi.org/10.6103/SHARE.w4.500
2. Börsch-Supan, A., Brandt, M., Hunkler, C., Kneip, T., Korbmacher, J., Malter, F., Schaan, B., Stuck, S., Zuber, S.: Data resource profile: the survey of health, ageing and retirement in Europe (SHARE). Int. J. Epidemiol. **42**, 992–1001 (2013)
3. Börsch-Supan, A., Brandt, M., Litwin, H., Weber, G. (eds.): Active Ageing and Solidarity Between Generations in Europe: First Results from SHARE After the Economic Crisis. De Gruyter, Berlin (2013)
4. Christelis, D., Jappelli, T., Padula, M.: Wealth and portfolio composition. In: Börsch-Supan, A., Brugiavini, A., Jürges, H., Mackenbach, J., Siegrist, J., Weber, G. (eds.) Health, Ageing and Retirement in Europe. First Results from the Survey on Health, Ageing and Retirement in Europe, pp. 310–317. MEA, Mannheim (2005)
5. Christelis, D., Jappelli, T., Paccagnella, O., Weber, G.: Income, wealth and financial fragility in Europe. J. Eur. Soc. Policy **19**, 359–376 (2009)
6. Fitzgerald, J., Gottschalk, P., Moffitt, R.: An analysis of sample attrition in panel data: the Michigan panel study of income dynamics. J. Hum. Resour. **33**, 251–299 (1998)
7. Grilli, L., Rampichini, C.: Selection bias in linear mixed models. Metron **68**, 309–329 (2010)
8. Guiso, L., Jappelli, T., Terlizzese, D.: Income risk, borrowing constraints, and portfolio choice. Am. Econ. Rev. **86**, 158–172 (1996)
9. Guiso, L., Haliassos, M., Jappelli, T. (eds.): Household Portfolios. MIT Press, Cambridge, MA (2002)
10. Heckman, J.: Sample selection bias as a specification error. Econometrica **47**, 153–162 (1979)
11. Hurd, M., Shoven, J.B.: The economic status of the elderly. In: Bodie, Z., Shoven, J.B. (eds.) Financial Aspects of the United States Pension System, pp. 359–398. University of Chicago Press, USA (1983)
12. Malter, F., Börsch-Supan, A. (eds.): SHARE Wave 4: Innovations & Methodology. MEA, Max Planck Institute for Social Law and Social Policy, Munich (2013)
13. McCarthy, D.: Household portfolio allocation: a review of the literature. Prepared for presentation at the Tokyo, Japan, February 2004 International Forum organized by the ESRI (2004)
14. Rabe-Hesketh, S., Skrondal, A., Pickles, A.: Multilevel Selection Models using *gllamm*. Stata User Group Meeting in Maastricht. http://fmwww.bc.edu/RePEc/dsug2002/select.pdf (2002)
15. Rabe-Hesketh, S., Skrondal, A., Pickles, A.: GLLAMM manual. U.C. Berkeley Division of Biostatistics Working Paper Series, Working Paper 160. www.biostat.jhsph.edu/~fdominic/teaching/bio656/software/gllamm.manual.pdf (2004)

# Bifurcation and Sunspots in Continuous Time Optimal Model with Externalities

**Beatrice Venturi and Alessandro Pirisinu**

**Abstract** In this chapter, we construct sunspot equilibria in a class of continuous time optimal control models with externalities. The model possesses stochastic characteristics which arise from indeterminate equilibrium and cycles (Hopf cycles) closed to the steady state. The model undergoes to Hopf bifurcations in some parameters set. We construct a stationary sunspot equilibrium near the closed orbit. We show that the stochastic approach suggests a way out from the poverty environments trap in a natural resource optimal model.

**Keywords** Multiple steady states · Sunspots · Oscillating solutions

## 1 Introduction

The standard approach of economic financial dynamical market is generally involved in the analysis of the conditions which determine convergence toward a (unique) stationary point along a (determinate) saddle path. This saddle path convergence has been considered, for a long time, as the "natural outcome of these models". Nevertheless, as soon as some form of market imperfection, e.g., an "externality" or an exogenous factor is introduced, such nice result may vanish and either a continuum of distinct equilibrium path or asymptotic convergence of certain classes of solutions to limit sets which are not stationary points, or both, can be established. An externality is the cost or benefit that affects a party who did not choose to incur that cost or benefit. An externality arises whenever the production/consumption of a good has an impact on the production/consumption of another good, and this effect is not measured by the price. A typical example of negative externality is the manufacturing activity

B. Venturi (✉) · A. Pirisinu
Department of Business and Economics, University of Cagliari, Cagliari, Italy
e-mail: venturi@unica.it

A. Pirisinu
e-mail: apirisinu@unica.it



F. Mola et al. (eds.), *Classification, (Big) Data Analysis and Statistical Learning*, Studies in Classification, Data Analysis, and Knowledge Organization,
https://doi.org/10.1007/978-3-319-55708-3_26

that causes air pollution, imposing health and clean-up costs on the whole society, whereas a positive externality is the creation of a new freeway which reduces travel times. So, unregulated markets in goods or services with significant externalities generate prices that do not reflect the full social cost or benefit of their transactions; such markets are therefore inefficient.

Following [3] as implications of this theoretical analysis, we construct cycles (Hopf cycles) and sunspots equilibria in a general class two-sector model. We are able to show that optimal control model with externalities possesses stochastic characteristics which arise from indeterminate equilibrium closed to steady state and cycles (Hopf cycles) (see [3, 8, 11]). In fact, by varying some characteristic parameter, the asymptotic behavior of the system (i.e., the behavior observed for very large times) can be of different type. The change in the asymptotic behavior that occurs due to the variations of a given parameter takes the name of bifurcation.

In the study of dynamical systems, the Hopf bifurcation occurs when, varying a certain control parameter, an equilibrium point changes its own stability in correspondence of the creation of a limit cycle (either attractive or repulsive) (see [5, 7]). The concept of sunspot equilibrium, as it is understood today, is the result of a long process of study and research. As reported in recent papers, the occurrence of the sunspots is compatible with the individual optimization, with self-fulfilling expectations and the compensations of competitive markets [1, 4, 9].

The sunspot activity is not economically significant if the following assumptions are made: strong rational expectations; completeness of markets; and absence of restrictions on the participation in the markets.

It also has to be considered that, in the macroeconomic theoretical debate, the research on sunspot was born as a reaction to the conclusions of Lucas and the new classical macroeconomics thesis related to the neutrality of money and the irrelevance of the economic policy. The theory of sunspot equilibria shows that there is room for the intervention of the State in the economic system which, eliminating restrictions to the participation in the market and other limiting conditions, enables to shift from dynamic Pareto-optimal to Pareto-optimal situations in the traditional sense. In other words, sunspot equilibria are a microeconomic way to highlight that there are macroeconomic equilibria of underemployment and, therefore, the government can intervene with expansive policies.

As an application of this general class of models, we consider a natural disposal resource model [2, 11]. In our formulation, the stochastic approach suggests a way out from the cycle trap (the poverty environment trap, see [9]). The paper develops as follows: the next sections analyze the general economic model and an application; the fourth section introduces stochastic dynamics related to the existence of sunspots in the economic model; the fifth section illustrates some simulations of the model; and the last section is devoted to display the results and the economic implications of existence of sunspots equilibrium in a natural resource system with externalities.

## 2 The Deterministic Economic General Model

The model deals with the maximization of an objective function (see [6])

$$\max_{c(t),u(t)} \int_0^{\infty} U(c)e^{-\rho t}dt \tag{1}$$

subject to

$$\dot{k} = A((r(t)^{\alpha_h}u(t)^{\alpha_u})(v(t)^{\alpha_v}k(t)^{\alpha_k})\widehat{r}(t)^{\alpha_{\widehat{r}}}k(t)^{\alpha_{\widehat{k}}} - \tau_k k(t) - c(t)$$
$$\dot{r} = B((r(t)^{\beta_r}(1-u(t)^{\beta_u}))((1-v(t)^{\beta_v}k(t)^{\beta_k})\widehat{r}(t)^{\beta_{\widehat{r}}}k(t)^{\beta_{\widehat{k}}} - \tau_r r(t)$$

$$k(0) = k_0$$

$$r(0) = r_0,$$

where

$$U(c) = \frac{c^{1-\sigma} - 1}{1-\sigma} \tag{2}$$

is a standard utility function, $c$ is per capita consumption, $\rho$ is a positive discount factor, and $\sigma$ is the inverse of the inter-temporal elasticity of substitution. The constraints are two equations related to the growth process of the analyzed economic system. Notation is as follows: $c$ is per capita consumption; $k$ is a physical capital and $r$ a natural resource (see [2, 8] the works of Lucas and Romer, where $r$ is the human capital or the knowledge, respectively); individuals have a fixed endowment of time, normalized to unity at each point in time, which is allocated to physical and to the other capital sector (natural resource); $\alpha_k$ and $\alpha_r$ being the private share of physical and the other capital sector in the output sector, $\beta_k$ and $\beta_r$ being the corresponding shares in the second sector, $u$ and $v$ are the fraction of aggregate other sector and physical capital used in the final output sector at instant $t$ (and conversely, $(1-u)$ and $(1-v)$ are the fractions used in the second sector); $A$ and $B$ are the levels of the technology in each sector; $\tau$ is a discount factor; $\alpha_{\widehat{k}}$ is a positive externality parameter in the production of physical capital; and $\alpha_{\widehat{r}}$ is a positive externality parameter in the production of the second sector. The equalities $\alpha_k + \alpha_r = 1$ and $\beta_k + \beta_r = 1$ ensure that there are constant returns to scale at the private level. At the social level, however, there may be increasing, constant, or decreasing returns depending on the signs of the externality parameters. All other parameters $\pi = (\alpha_k, \alpha_{\widehat{k}}, \alpha_r, \alpha_{\widehat{r}}, \beta_k, \beta_{\widehat{k}}, \beta_r, \beta_{\widehat{r}}, \sigma, \gamma, \delta, \rho)$ live inside the following set: $\Pi \subset (0,1) \times (0,1) \times (0,1) \times (0,1) \times (0,1) \times (0,1) \times (0,1) \times (0,1) \times \mathbb{R}^4_+$.

The representative agent's problem (1)–(2) is solved by defining the current value Hamiltonian:

$$H = \frac{c^{1-\sigma} - 1}{1-\sigma} + \lambda_1(A((r(t)^{\alpha_r} u(t)^{\alpha_u})(v(t)^{\alpha_v} k(t)^{\alpha_k})\widehat{r}(t)^{\alpha_{\widehat{r}}} k(t)^{\alpha_{\widehat{k}}} - \tau_k k(t) - c(t)) +$$
$$+ \lambda_2(B((r(t)^{\beta_r}(1-u(t)^{\beta_u}))((1-v(t)^{\beta_v} k(t)^{\beta_k})\widehat{r}(t)^{\beta_{\widehat{r}}} k(t)^{\beta_{\widehat{k}}} - \tau_r r(t)),$$

where $\lambda_1$ and $\lambda_2$ are co-state variables which can be interpreted as shadow prices of the accumulation. The solution candidate comes from the first-order necessary conditions (for an interior solution) obtained by means of the Pontryagin maximum principle with the usual transversality condition:

$$\lim_{t\to\infty} \left[e^{-\rho t}(\lambda_1 k + \lambda_2 r)\right] = 0. \tag{3}$$

We consider only the competitive equilibrium solution (as well known, it follows from the presence of the externality that the competitive solution differs from the planner's solution). After eliminating $v(t)$, the rest of the first-order conditions and accumulation constraints entail four first-order nonlinear differential equations in four variables: two controls ($c$ and $u$) and two states ($k$ and $r$). The solution of this autonomous system is called a balanced growth path (BGP) if it entails a set of functions of time solving the optimal control problem (1)–(3) such that $k$, $r$, and $c$ grow at a constant rate and $u$ is constant. With a change of variable in standard way (since $k$, $r$ and $c$ grow at a constant rate and $u$ is a constant in the BGP), we transform a system of four first-order ordinary differential equations in $c$, $u$, $k$, and $r$ into a system of three first-order ordinary differential equations with two non-predetermined variables (the control variables) and one predetermined (a linear combination of the state variables).

Setting $A = B = 1$ and: $x_1 = kr^{\frac{\alpha_{\widehat{r}}}{(\alpha_{\widehat{r}}-1)}}$; $x_2 = u$; $x_3 = \frac{c}{k}$, we get: $\dot{x}_1 = \phi_1(x_1, x_2, x_3)$; $\dot{x}_2 = \phi_2(x_1, x_2, x_3)$; $\dot{x}_3 = \phi_3(x_1, x_2, x_3)$. In vectorial form: $(\dot{x}_1, \dot{x}_2, \dot{x}_3)^T = \Phi(x_1, x_2, x_3)$, where the $\Phi \in R^3$, are continuous and derivable complicated nonlinear functions, which depend on the parameters $(\alpha_k, \alpha_{\widehat{k}}, \alpha_r, \alpha_{\widehat{r}}, \beta_k, \beta_{\widehat{k}}, \beta_r, \beta_{\widehat{r}}, \sigma, \gamma, \delta, \rho)$ of the model, and $\Phi : UxR^3 \longrightarrow R^3$ with $U \subset R$, an open subset, and $i = 1, 2, 3$.

## 3 Application: Natural Resource System

Our natural disposal resource system includes two non-predetermined variables $x_1$ and $x_2$ and one predetermined variable $x_3$:

$$\begin{aligned}\dot{x}_1 &= (-\tfrac{\rho}{\sigma})x_1 + (\tfrac{\beta-\sigma}{\sigma})x_1 - x_1^2\\ \dot{x}_2 &= (\tfrac{\gamma\delta}{\beta})(1-x_2)x_2 + x_1x_2\\ \dot{x}_3 &= (\tfrac{\gamma\delta}{\beta})((1-x_2)x_3 + (\beta-1)x_3^2,\end{aligned} \tag{4}$$

where $x_1 = \frac{c}{k}$; $x_2 = nr$; $x_3 = \frac{y}{k}$. We found eight steady-state values:
$P_1^*$ $(0,0,0)$; $P_2^*$ $\left(0, 0, \left(\frac{\gamma\delta}{\beta(1-\beta)}\right)\right)$; $P_3^*$ $(0,1,0)$ (*double sol.*); $P_4^*$ $\left(\frac{\rho}{\sigma}, 0, 0\right)$;
$P_5^*$ $\left(\frac{1}{\sigma}\left(\rho - \frac{\gamma\delta(\beta-\sigma)}{\beta(1-\beta)}\right), 0, \left(\frac{\gamma\delta}{\beta(1-\beta)}\right)\right)$; $P_6^*$ $\left(\frac{\rho}{\sigma}, 1 - \frac{\beta\rho}{\sigma\gamma\delta}, 0\right)$;
$P_7^*$ $\left(\frac{\rho(1-\beta)}{\beta(1-\sigma)}, 1 - \frac{\rho(1-\beta)}{\gamma\delta(1-\sigma)}, \frac{\rho}{\beta(1-\sigma)}\right)$.

It is well known that many theoretical results relating to the system depend upon the eigenvalues of the Jacobian matrix evaluated at the stationary point $P_i^*$ with $i = 1, 2, 3, 4, 5, 6, 7$ for some values of the parameters.

## 4 Stochastic Dynamic

We remember that a probability space is a triple $(\Omega, B_{R^2}, P_{R^2})$ where $\Omega$ denotes the space of events, and $B$ is the set of possible outcomes of a random process; $B$ is a family of subsets of $\Omega$ that, from a mathematical point of view, represents a $\sigma$ -algebra. The $\sigma$-algebra can be interpreted as information (on the properties of the events). We add a "noise" (a Wiener process) in the equations related to the control variables of the optimal choice problem. Finally, we get the following stochastic equation:

$$\begin{aligned}dx_{1t} &= \phi_1(x_{1t}, x_{2t}, x_{3t})dt + sd\varepsilon_t\\ dx_{2t} &= \phi_2(x_{1t}, x_{2t}, x_{3t})dt + d \ni_t\\ x_{3t}dt &= \phi_3(x_{1t}, x_{2t}, x_{3t})dt.\end{aligned} \tag{5}$$

Let $s_t(\omega) = (\omega, t)$ be a random variable irrelevant to fundamental characteristic of the optimal economy; it means that does not affect preferences, technology, and endowment (i.e., sunspot). We assume that a set of sunspot variable $\{s_t(\omega)\}_{t\geq 0}$ is generated by a two-state continuous time Markov process with stationary transition probabilities and that $s_t : \Omega \longrightarrow \{1, 2\}$ for each $t \geq 0$. Let $[\{s_t(\omega)\}_{t\geq 0}, (\Omega, B_{R^2}, P_{R^2})]$ be a continuous time stochastic process, where $\omega \in \Omega$, $B_\Omega$ is a $\sigma$-field in $\Omega$, and $P_\Omega$ is a probability measure. We assume that the probability space is a complete measure space and the stochastic process is separable.

Let $(R^2_{++}, B_{R^2_{++}}, P_{R^2_{++}})$ be a probability space on the open subset $R^2_{++}$ of $R^3$ where $B_{R^2_{++}}$ denotes the Borel $\sigma$-field in $R^2_{++}$. Let $(\Phi, B, P)$ be the product probability space of $(R^2_{++}, B_{R^2_{++}}, P_{R^2_{++}})$ and $(\Omega, B_\Omega, P_\Omega)$. That is $(R^2_{++} \times \Omega, B_{R^2_{++}} \times B_\Omega, P_{R^2_{++}} \times P_\Omega)$. Let $(\Phi, B^*, P^*)$ be the completion of $(\Phi, B, P)$. Let $(x_0^1, x_0^2, x_0^3)$ be the value of our model at the time $t = 0$. We denote a point $(x_0^1, x_0^2, x_0^3, \omega)$ in $\Phi$ as $\varphi$: in other words, $\varphi = (x_0^1, x_0^2, x_0^3, \omega)$. Let $B_t = B(x_0^1, x_0^2, x_0^3, s_s, s \leq t)$ the smallest $\sigma$-field of $\varphi$ respect to which $(x_0^1, x_0^2, x_0^3, s_s, s \leq t)$ are measurable. Let $B_t^* = B^*(x_0^1,$

$x_0^2, x_0^3, s_s, s \leq t$) be the $\sigma$-field of $\varphi$ sets which are either $B_t$ sets or which differ from $B_t$ sets by sets of probability zero. Let $E_t$ the conditional expectation operator relative to $B_t^*$.

The following equation is a first-order condition of some inter-temporal optimization problem with market equilibrium conditions incorporated:

$$(E_t(d\dot{x}_1/dt), E_t(d\dot{x}_2/dt), \dot{x}_3) = \Phi(x_1(\varphi), x_2(\varphi), x_3(\varphi)) \tag{6}$$

where $(x_0^1(\varphi), x_0^2(\varphi), x_0^3(\varphi)) = (x_0^1, x_0^2, x_0^3)$, and $\frac{dx_{1t}}{dt}, \frac{dx_{2t}}{dt}, \frac{dx_{3t}}{dt}$ are defined as $\frac{dx_t^i}{dt} = \lim_{h \to 0^+} \frac{(x_{t+h}^i - x_t^i)}{h} (i = 1, 2, 3)$ if the limit exists. Our system (4) includes one predetermined variable $x_3$ and two non-predetermined variables $x_1$ and $x_2$.

## 4.1 Simulations

The deterministic equilibrium dynamic of the natural resource model (4) has a family of periodic orbits $\Gamma_{\sigma_c}$ emerging from one steady state, with $\Gamma_{\sigma_c}$ in the center manifold (a two-dimensional invariant manifold in $R_{++}^2$). For some set of parameters in the model (see [2]), there exists a sunspot equilibrium whose support is located in the bounded region enclosed by the periodic orbit $\Gamma_{\sigma_c}$. Each sample path of the sunspot equilibrium does not converge to any specific point and continues to fluctuate without decaying asymptotically (Fig. 1).

**Theorem 1** *A family of Hopf bifurcations emerges around the steady-state $P_7^*$ of the system (4), for $\sigma = \sigma_c^*$ and $\sigma = \sigma_c^{**}$.*

We consider some numerical simulations: in particular, we consider the following two sets of parameters (see [2]):

(*a*) $\rho = 0.002,\ \beta = 0.66, \gamma = 2, \delta = 0.04, \sigma_c^* = 0.66$;

(*b*) $\rho = 0.002, \beta = 0.66, \gamma = 2, \delta = 0.04, \sigma_c^{**} = 0.975$.

**Theorem 2** *A sunspot equilibrium (SE) is a stochastic process $\{(x_0^1(\varphi), x_0^2(\varphi), x_0^3(\varphi))\}_{t \geq 0}$ with a compact support (the periodic orbits $\Gamma_{\sigma_c}$) such that it is a solution of a stochastic differential equation (5).*

*Proof* It follows directly from the main Theorem in Nishimura, T. Shigoka [8].

We evaluate the growth rate of the economy and we get $\mu = \rho - \delta(1 - x_2^*)$. We have two cases:

(a) if $\sigma_c^* = \beta$, then: $\mu = \rho - \delta\left(1 - \frac{\rho}{\beta}\right)$; substituting the values of the parameters, we get $\mu = f(\delta) = 0.002 - 0.997\delta$.

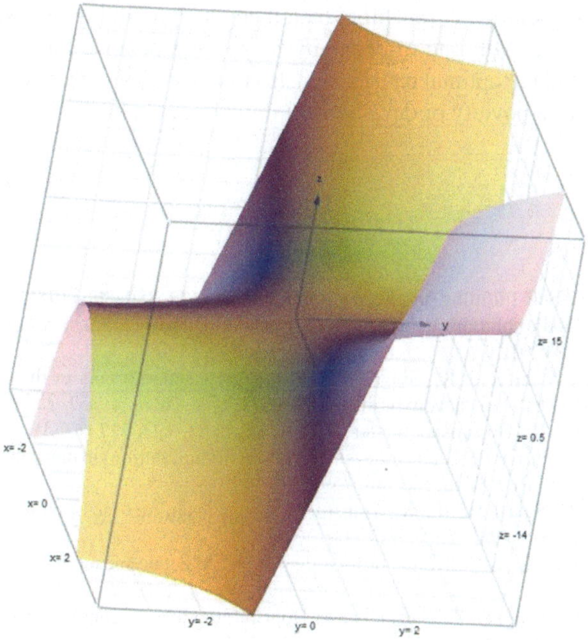


**Fig. 1** Growth rate of the economy as a function of both $(\delta, \gamma)$

(b) if $\sigma_c^{**} = \frac{\gamma\delta - \rho(1-\beta)}{\gamma\delta}$, then: $\mu = \rho - \delta\left(1 - \frac{\gamma\delta}{\beta}\right)$; substituting the values of the parameters, we get $\mu = f(\delta) = 0.002 - \delta + 3.03\delta^2$.

If we consider $\mu$ as a function of both $(\delta, \gamma)$, we get $\mu = f(\delta, \gamma) = 0.002 - \delta\left(1 - \frac{\gamma\delta}{0.66}\right)$.

## 5 Conclusions

One strand of endogenous growth theory assumes that investment in capital shows positive external effects; among many, authors like [6] have shown that the investment share is a robust variable in explaining economic growth. It is important to underline that, in case of endogenous growth, the balanced growth rate crucially depends on the marginal product of physical capital, which varies positively with the stock of knowledge capital. Thus, countries with a smaller stock of purely physical capital tend to have higher growth rates than countries with a larger stock, when the level of knowledge capital plays its important role.

For some parameter value, due to pessimistic self-fulfilling expectations, sunspot equilibria may exist in some neighborhood of the steady state. If the periodic orbit emerging from the steady state is supercritical, there is no way out [9]. If the cycle is subcritical the solution is repelling, then there is a possibility of a way out of the orbit (in fact the growth rate of economy $\mu$ became positive for low level of the externality $\gamma$) and the optimal path can reach another steady state. Such situation can be understood as a poverty or development trap respectively.

## References


1. Azariadis, C.: Self-fulfilling prophecies. J. Econ. Theor. **25**, 380–396 (1981)
2. Bella, G.: Periodic solutions in the dynamics of an optimal resource extraction model. Environ. Econ. **1**, 49–58 (2010)
3. Benhabib, J., Nishimura, K., Shigoka, T.: Bifurcation and sunspots in the continuous time equilibrium model with capacity utilization. Int. J. Econ. Theor. **4**, 337–355 (2008)
4. Cass, D., Shell, K.: Do sunspot matter? J. Polit. Econ. **91**, 193–227 (1983)
5. Mattana, P., Venturi, B.: Existence and stability of periodic solutions in the dynamics of endogenous growth. Int. Rev. Econ. Bus. **46**, 259–284 (1999)
6. Mulligan, C.B., Sala-I-Martin, X.: Transitional dynamics in two-sector models of endogenous growth. Q. J. Econ. **108**, 739–773 (1993)
7. Neri, U., Venturi, B.: Stability and bifurcations in IS-LM economic models. Int. Rev. Econ. **54**, 53–65 (2007)
8. Nishimura, K., Shigoka, T.: Sunspots and Hopf bifurcations in continuous time endogenous growth models. Int. J. Econ. Theor. **2**, 199–216 (2006)
9. Slobodyan, S.: Sunspot fluctuations: a way out of a development trap? CERGE-EI Working Paper Series No. 175, available at SSRN: https://ssrn.com/abstract=1514542 or https://doi.org/10.2139/ssrn.1514542, (2001)
10. Venturi, B.: Chaotic solutions in non linear economic–financial models. Chaotic Model. Simul. (CMSIM) **3**, 233–254 (2014)
11. Wirl, F.: Sustainable growth, renewable resources and pollution: thresholds and cycles. J. Econ. Dyn. **28**, 1149–1157 (2004)

# Erratum to: Big Data Meet Pharmaceutical Industry: An Application on Social Media Data

**Caterina Liberati and Paolo Mariani**

## Erratum to: Chapter "Big Data Meet Pharmaceutical Industry: An Application on Social Media Data" in: F. Mola et al. (eds.), *Classification, (Big) Data Analysis and Statistical Learning*, Studies in Classification, Data Analysis, and Knowledge Organization, https://doi.org/10.1007/978-3-319-55708-3_3

In the original version of the book, the following correction has been incorporated:

In chapter "Big Data Meet Pharmaceutical Industry: An Application on Social Media Data", Fig. 1 has been replaced with a new figure.

---

The updated online version of this chapter can be found at
https://doi.org/10.1007/978-3-319-55708-3_3



F. Mola et al. (eds.), *Classification, (Big) Data Analysis and Statistical Learning*,
Studies in Classification, Data Analysis, and Knowledge Organization,
https://doi.org/10.1007/978-3-319-55708-3_27